BURT P

SMALL GASOLINE ENGINES

FOURTH EDITION

GEORGE E. STEPHENSON

DELMAR PUBLISHERS INC.®

COVER PHOTOS:
Courtesy of Tecumseh Products Company;
Lauson-Power Products Engine Division
Grafton, Wisconsin

Delmar Staff
Administrative Editor: Mark W. Huth
Production Editor: Ruth Saur

For information, address Delmar Publishers Inc.
2 Computer Drive West, Box 15–015
Albany, New York 12212

Printed in the United States of America
Published simultaneously in Canada
by Nelson Canada,
A division of International Thomson Limited

10   9   8   7

Library of Congress Cataloging in Publication Data

Stephenson, George E.
   Small gasoline engines.

   Includes index.
   1. Internal combustion engines, Spark ignition.
I. Title.
TJ790.S73   1984   621.43′4   83-18943
ISBN 0-8273-2242-9

CONTENTS

Unit 5 Lubrication. 87

Friction Bearings; Antifriction Bearings; Quality Designation of Oil and SAE Number; Quality Designation by the American Petroleum Institute (API); Viscosity Classification by the Society of Automotive Engineers; Motor Oil Additives; *Lubrication of Four-cycle Engines;* Simple Splash System; Constant-level Splash System; Ejection Pump System; Barrel-type Pump System; Full-pressure Lubrication System; *Lubricating Cylinder Walls; Blow-by; Crankcase Breathers; Oil Changes; Lubrication of Two-cycle Engines; Additional Lubrication Points*

Unit 6 Cooling Systems 107

Air-cooling System; Care of the Air-cooling System; *Water-cooling System; Water Cooling the Outboard Engine*

Unit 7 Ignition Systems 119

Electron Theory; Units of Electrical Measurement; Ampere; Volt; Ohm; *Ohm's Law; Magnetism; Permanent Magnets; High-tension Coil (Primary and Secondary Windings); Laminated Iron Core; Breaker Points; Breaker Cam; Condenser; Spark Plug Cable; Spark Plug; The Complete Magneto Cycle; Spark Advance; Manual Spark Advance; Automatic Spark Advance; Impulse Coupling; Magneto Ignition for Multicylinder Engines; Engine Timing; Solid State Ignition; Spark Plug; Battery Ignition*

Unit 8 Routine Care and Maintenance, and Winter Storage 149

Routine Care and Maintenance; Oil Supply; Cooling System; Spark Plugs; Air Cleaner; Oil-type Air Cleaners; Dry-type Air Cleaners; *Exhaust Ports (Two-cycle);*

Winter Storage and Care; Returning the Engine to Service; Special Considerations for Outboard Motors; Winter Storage; Salt Water Operation; Use in Freezing Temperatures

Laboratory Experiences

Unit 9   Troubleshooting, Tune-up, and Reconditioning . . . .

Engine Tune-up; Reconditioning; Tools and Equipment; Basic Steps; *Flywheel; Magneto Parts; Engine Timing; Cylinder Head; Valves; Piston and Rod Assembly; Cylinder; Piston; Piston Pin; Piston Rings; Crankshaft; Camshaft and Tappets*

Laboratory Experiences

Unit 10   Horsepower, Specifications, and Buying Considerations . . . .

Factors Affecting Horsepower; Estimating Horsepower; *Specifications; Buying Considerations*

Unit 11   Mechanical Transmission of Power . . . .

Mechanical Transmission of Power; Lever; Wheel and Axle; Inclined Plane; Pulley; *Gears; The Power Train;* Clutch; Transmission; Universal Joint; Differential; Transaxles

Laboratory Experience

PREFACE

SMALL GASOLINE ENGINES explores a power source that enjoys immense success and widespread use throughout the country. Millions of people use small gasoline engines in recreation and work activities. Lawn mowers, chain saws, outboard motors, portable generators, portable pumps, motor bikes and motor cycles are some of the common applications for small engines.

For those who intend to qualify as mechanics of small gasoline engines and for all who operate and maintain these power sources, SMALL GASOLINE ENGINES provides basic instruction.

Now in its fourth edition, SMALL GASOLINE ENGINES continues to keep pace with technological advances. The following topics summarize changes made in this latest edition.

- New unit on the mechanical transmission of power
- Expanded coverage on troubleshooting, tune-up and reconditioning
- Expanded appendix
- Battery ignition for small engines
- Common valve problems
- Second color added to increase clarity

The popular format of SMALL GASOLINE ENGINES has been maintained. Each unit is treated as a single coherent block of instruction and is followed with:

- Review questions
- Class discussion topics
- Class demonstration topics
- Laboratory experiences

The Laboratory Experiences are of special significance since they are designed to guide the student in the procedures of assembly and disassembly. The Experiences direct the work in an orderly fashion while the student investigates the workings of the small gasoline engine. The Laboratory Experiences are closely correlated to the appropriate units so that the combination of theory and practice provides for a logical and unified instructional program. The procedures given in the Laboratory Experiences are fundamental and apply to all makes of small gasoline engines.

The many illustrations and clearly written information makes SMALL GASOLINE ENGINES worthwhile for all those with an interest in engines.

Reviewers for the Fourth Edition

William Tedford, Instructor, Heyers Vocational Technical Center, Saratoga Springs, NY

Michael Burt, Instructor, S.C.T. BOCES, Elmira, NY

Kenneth Parnapy, Instructor, North Franklin Education Center, Malone, NY

Jerry D. Newlin, Instructor, Industrial Education Department, Atwood-Hammond High School, Atwood, IL

John Nelson, Instructor, Central Piedmont Community College, Charlotte, NC

TO THE STUDENT

The fourth edition of SMALL GASOLINE ENGINES has been made into a hardcover text. Therefore, students should check with the instructor to receive directions for answering review questions and using the work record box for Laboratory Experiences.

NOTICE TO THE READER

UNIT 1

Work, Energy, and Power

OBJECTIVES   After completing this unit, the student should be able to:
- Define the terms work, energy, power and foot-pounds.
- State and use the formulas for work, efficiency, power and horsepower.
- State the difference between kinetic and potential energy.
- State the law of the conservation of energy.
- Name the forms that energy can take.
- Explain how friction affects the efficiency of machines.

Internal combustion engines are the work-horses of our industrial society. These engines are often taken for granted even though they have been in existence for less than 200 years. Before the era of engines, energy was derived mainly from the wind, water power, and draft animals.

James Watt developed the first practical steam engine in 1765. The engine burned fuel outside the engine in order to produce the steam needed to run the engine. These first steam engines were called external combustion engines. They were large and heavy and suited for use at factory sites, in coal mines, on ships and locomotives. The development of external combustion engines paved the way for the modern internal combustion engine which burns fuel inside the engine itself.

Early pioneers in internal combustion engines, such as Christian Huygens and J.J.E. Lenoir, developed workable engines. By 1877, Nikolas A. Otto had built a successful internal combustion engine. It was large and heavy by today's standards but the principles on which it operated are still used today. Other pioneers such as Gottleib Daimler soon developed lighter engines suitable for use on trucks and automobiles.

Continued improvements in engine design have made the lightweight small gasoline engine a reality for almost every industry and household in the country. More than 70 million small engines are in use today powering lawn mowers, chain saws, outboard motors and many other machines that are used for work and recreation.

A basic understanding of small gasoline engines can benefit almost every person. Before starting a detailed study of the elements of small gasoline engines, it is necessary to review some of the terms which are basic to an understanding of mechanical power. What is work? What is energy? What is power? Are these terms all the same and if not how do they differ? It is the purpose of this unit to define and explain the meaning of these terms.

WORK

Work is a scientific term as well as an everyday term. To many people, work means engaging in an occupation. One person may work hard in an office while another person may work hard carrying bricks on a construction job. Both individuals may come home tired but, scientifically, only the one carrying bricks has done much work.

Work, in the scientific sense, involves the moving of things by applying force. *Work* is defined as applying a force through a distance. In other words, work is motion caused by applying a force.

Measurement of Work

The common unit for measuring work is the *foot-pound*. The amount of work required to lift one pound, one foot is expressed as one foot-pound. A person carrying 50 pounds of bricks up a 10-foot ladder does 500 foot-pounds of work. Stated as a formula:

$$\text{Work} = \text{Force} \times \text{Distance}$$
$$(\text{Ft - lb}) = (\text{Pound}) \times (\text{Feet})$$

Figure 1-1 shows how to calculate the work involved in lifting a 20-pound weight 5 feet.

$$\text{Work} = \text{Force} \times \text{Distance}$$
$$\text{Work} = 20 \text{ lb} \times 5 \text{ ft} = 100 \text{ ft - lb}$$

By scientific definition, the person who struggles to move a boulder but fails to budge it is not doing work because there is no "distance moved." Engines can do work because they are capable of applying force to move machines, loads, implements, and so forth through distances.

ENERGY

Energy is the ability to do work. The energy stored in the human body, for instance, supplies the body with a potential to do work. A person can work for a long period of time on stored energy before needing to refuel with food. The human body is a good example of potential energy.

Potential Energy

Potential energy is the energy a body has due to its position, its condition, or its chem-

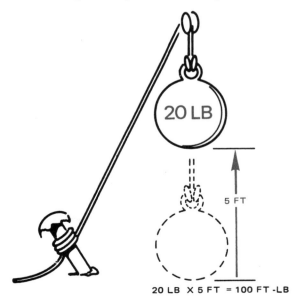

20 LB X 5 FT = 100 FT -LB

Fig. 1-1 A twenty-pound weight lifted five feet in the air.

ical state. The following are some examples of potential energy: water at the top of a waterfall (position); a tightly wound watch spring (condition); fuels such as gasoline or coal, and foodstuffs to be burned by the body (chemical state).

Stored or potential energy is measured in the same way as work, in foot-pounds. How much potential energy is there when a 20–pound weight is lifted 5 feet?

Potential Energy = Force × Distance
Potential Energy = (weight) × (height)
Potential Energy = 20 lb × 5 ft
Potential Energy = 100 ft-lb

With the weight at the 5-foot height, there is 100 foot-pounds of potential energy.

Kinetic Energy

Kinetic energy is the energy of motion. Some examples are the energy of a thrown ball, the energy of water falling over a dam, and the energy of a speeding automobile. Kinetic energy is, in effect, released potential energy.

Consider a thrown ball, the work of throwing it, and what becomes of it. If a ball is thrown by applying a force of 8 pounds through 6 feet, how much work is done?

Work = Force × Distance
Work = 8 lb × 6 ft
Work = 48 ft-lb

What becomes of this energy? It exists as energy of motion of the ball — kinetic energy. The ball in flight has 48 foot-pounds of energy which it delivers to whatever it hits. Refer back to the example of potential energy. When the 20-pound weight which rests 5 feet above the floor is allowed to fall, its potential energy of 100 foot-pounds becomes 100 foot-pounds of kinetic energy, figure 1-2.

Energy can change form but it cannot be destroyed. This is the *law of the conservation of energy*. Energy can take the form of light, sound, heat, motion, and electricity. Consider the potential chemical energy of gasoline in an engine. Upon ignition, its energy is converted mainly into heat energy. The heat energy is used to push the pistons down and is the kinetic energy of the rotating crankshaft.

Consider, again, the 20-pound weight at a 5-foot height. When the weight falls to the floor, its energy is not lost; it is transformed into other types of energy. The floor becomes slightly warmer at the point of impact due to the creation of heat energy. A sound can be heard at the moment of impact, indicating the presence of sound energy. The floor gives slightly, thus absorbing some energy.

The efficiency of machines that transform energy is not 100 percent. Some energy is wasted in moving the machine parts to overcome friction which is inherent in all machines.

20 LB × 5 FT = 100 FT -LB

Fig. 1-2 A twenty-pound weight dropped to the floor from a height of five feet delivers 100 foot-pounds of kinetic energy.

The *efficiency* of a machine is the ratio of the work done by the machine to the work put into it.

Stated as a formula:

$$\text{Efficiency} = \frac{\text{Output}}{\text{Input}} \times 100$$

or

$$\text{Efficiency} = \frac{\text{Input} - \text{Losses}}{\text{Input}} \times 100$$

or

$$\text{Efficiency} = \frac{\text{Output}}{\text{Output} + \text{Losses}} \times 100$$

The figure 100 in these formulas is a way of converting the fraction to a percentage.

POWER

In everyday language the word *power* can mean different things; for example, political power is different than physical power. However, in the language of scientists and engineers, power refers to how fast work is done or how fast energy is transferred. *Power* is the rate of doing work or the rate of energy conversion.

How fast a machine can work is an important consideration for engineers and consumers. For example, buying gasoline is a purchase of potential energy. The size and efficiency of the engine determine how fast this energy can be converted into power. A 10-horsepower engine can work only half as fast as a 20-horsepower engine.

Measurement of Power

Power is measured in foot-pounds per second or in foot-pounds per minute. Stated as a formula:

$$\text{Power} = \frac{\text{Work}}{\text{Time}}$$

Referring again to the 20-pound weight that is lifted 5 feet, how much power is required to lift the weight if it takes 5 seconds?

$$\text{Work} = \text{Force} \times \text{Distance}$$
$$\text{Work} = 20 \text{ lb} \times 5 \text{ ft} = 100 \text{ ft-lb}$$

$$\text{Power} = \frac{\text{Work}}{\text{Time}}$$

$$\text{Power} = \frac{100 \text{ ft-lb}}{5 \text{ seconds}}$$

$$\text{Power} = 20 \text{ ft-lb/second}$$

As another example, if an elevator lifts 3500 pounds a distance of 40 feet and it takes 25 seconds to do it, what is the rate of doing work?

$$\text{Work} = \text{Force} \times \text{Distance}$$
$$\text{Work} = 3500 \text{ lb} \times 40 \text{ ft} = 140,000 \text{ ft-lb}$$

$$\text{Power} = \frac{\text{Work}}{\text{Time}}$$

$$\text{Power} = \frac{140,000 \text{ ft-lb}}{25 \text{ seconds}} = 5600 \text{ ft-lb/sec}$$

Power can also be expressed in foot-pounds per minute. If a pump needs 10 minutes to lift 5000 pounds of water 60 feet, it is doing 300,000 foot-pounds of work in 10 minutes, which is a rate of 30,000 foot-pounds per minute.

Horsepower

The power of most machinery is measured in horsepower. The term *horsepower* originated many years ago when James Watt, attempting to sell his new steam engines, had to rate the engines in comparison with the horses they were to replace. He found that an average horse working at a steady rate could do 550 foot-pounds of work per second, figure 1-3. This rate is the definition of *one horsepower*.

The formula for horsepower is:

$$\text{Horsepower} = \frac{\text{Work}}{\text{Time (in seconds)} \times 550}$$

If time in this formula is expressed in minutes, it is multiplied by 550×60 (seconds) or 33,000. The formula, then, may be expressed as:

$$\text{Horsepower} = \frac{\text{Work}}{\text{Time (in minutes)} \times 33,000}$$

What horsepower motor would the elevator previously referred to have?

$$\text{Horsepower} = \frac{\text{Work}}{\text{Time (in seconds)} \times 550}$$

$$\text{Horsepower} = \frac{140,000 \text{ ft-lb}}{25 \text{ seconds} \times 550}$$

$$\text{Horsepower} = 10 +$$

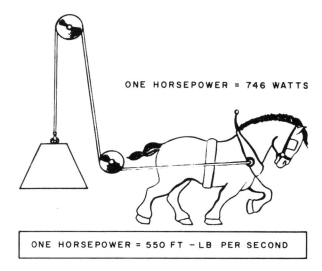

ONE HORSEPOWER = 746 WATTS

ONE HORSEPOWER = 550 FT – LB PER SECOND

Fig. 1-3 Horsepower

Fig. 1-4 Both outboard motor engines are entirely suitable for the job requirements. However, the larger horsepower engine can convert energy to work at a faster rate.

REVIEW QUESTIONS

1. Give the definition of work.

2. What is the unit of measurement for work?

3. A 100-pound boy climbs a 16-foot stairway. How much work has he done?

4. Define energy.

5. Explain the difference between kinetic and potential energy.

6. a. What is the law of the conservation of energy?
   b. Name five forms that energy can take.

7. Explain why a machine cannot be 100 percent efficient.

8. Define power.

9. What is the unit of measurement for power?

10. How much power is needed to lift a 100-pound bag of cement in 2 seconds onto a truck bed that is 3 feet high?

11. What is the formula for horsepower using time in seconds? Using time in minutes?

12. How much horsepower does an engine on a grain elevator deliver if it can load 440 pounds of corn into a 25-foot storage bin in 10 seconds?

CLASS DISCUSSION TOPICS

- Discuss the difference between scientific work and other kinds of work.

- Discuss and list potential and kinetic energy sources.

- Discuss the origin of the term *horsepower*.

- Discuss the energy changes involved when a gasoline engine is used with a portable generator to operate devices such as a radio, toaster, light bulb, etc.

- Discuss how energy changes form in the power plant that supplies your community with electrical energy.

- Discuss the difference in power involved when a small child works at moving a large pile of books across the room and when a large student does the same job. Do they do the same work? Are they equally powerful?

- Discuss the many applications of small gasoline engines.

CLASS DEMONSTRATION TOPICS

- Have a student lift a given weight a definite height and calculate the work.
- Demonstrate how the horsepower developed by a student can be calculated.
- Demonstrate the transformation of energy by striking and burning a match.
- Demonstrate potential and kinetic energy sources.
- Demonstrate the conservation of energy.
- Demonstrate loss due to friction and how it affects the efficiency of a machine.

UNIT 2

Safety

OBJECTIVES   After completing this unit, the student should be able to:
- Explain how personal attitudes toward safety affect other people.
- List rules for storing gasoline.
- Name safety precautions for gasoline use.
- List safety rules for operating power lawn mowers.
- Identify the proper fire extinguisher to be used with each class of fire.
- State how carbon monoxide is produced.
- List the effect that carbon monoxide has on the body.
- State rules for the safe use of hand tools.

The maintenance, operation, repair, and testing which are involved in a study of small gasoline engines requires that careful and constant emphasis be placed on safety. It is beyond the scope of this unit to deal with the particular safety considerations of each different engine and its vast number of applications. Instead, this unit deals with safety considerations in several common areas: safety attitudes, safety with gasoline, fires, carbon monoxide safety, safety with basic hand tools, safety with basic machines and appliances, and safety with small gasoline engines.

Specific safety information is included in later units as it applies. However, it is useful to review this unit before beginning laboratory experiences.

SAFETY ATTITUDES

Safety is important to every student, both as an individual and as a member of a community. The machine age brought many safety hazards. In transportation alone millions of power-driven vehicles presented a number of hazards. Industrial, construction,

8

and scientific employees work with the tools of power technology every day. There are also many hazards in the home — power appliances, household machines, fire, boiling liquids, electricity, and so forth. Considering these hazards, it is necessary that an emphasis on safety begin early in childhood and continue throughout life.

Accidents can be compiled into statistics. Accidents can be measured in terms of frequency, probability, medical costs, time lost from work, or in many other ways that serve a useful purpose in identifying hazards and in developing safety programs.

On an individual basis, the statistical approach loses all of its importance. How can a number or a dollar sign be put on the personal cost and human suffering that result when accidents occur? People do not like to think of the unpleasantness of an accident happening to them. Fortunately, most try to protect themselves and those close to them from hazardous situations and accidents.

In gasoline engine safety, people must consider both their own safety, and the safety of others. Each person's actions and safety attitudes do have an effect on other individuals. Sometimes the effect is slow and indirect, such as parents' carelessness becoming part of their children's safety standards. At other times, the effect on others is direct, such as a head-on collision on the highway caused by an irresponsible driver. The point is that in power safety, individuals are responsible for themselves as well as for the safety of others.

Students working with small gasoline engines should consider the safety aspect of every activity in which they engage. Most basic safety approaches to prime movers and machines are similar and the knowledge gained from one situation can be applied to another. Safety is never sacrificed for the sake of speed or expediency. It should have

priority over everything else. How a person operates a machine can reveal that person's entire safety code. It can also indicate personal qualities or deficiencies.

Learning safety by trial and error alone is not enough. The safety considerations of each situation must be studied. A defensive attitude of, "if this happens, what might the result be?" must be a part of a person's overall judgment. Some safety is common sense. However, many safety requirements must be learned from machinery instruction booklets and repair manuals. To ensure personal safety as well as safety for others, the operator must know specific information that applies to the particular prime mover and its uses.

SAFETY WITH GASOLINE

Understanding the nature of gasoline and its safe use is very important. Gasoline is used as a common item in many households. The presence of this liquid, which is more powerful than TNT, is readily accepted. Many people have become too casual about this volatile and explosive liquid.

Gasoline is around the home (in the garage, toolshed, or barn) largely because the use of small gasoline engines has become so widespread. Lawn mowers, garden tractors, snow throwers, go-carts, utility vehicles, motorbikes, outboard motors, and chain saws need gasoline for operation.

Unfortunately, accidents do happen. It is estimated that every year several hundred persons lose their lives due to accidents involving gasoline and other flammable liquids. In addition, several million dollars of property damage is caused by such accidents.

The volatility of gasoline is the characteristic that makes it such an ideal fuel. This same characteristic is also the prime danger. Gasoline can vaporize and explode in the atmosphere just as it can in the engine. A concentration of one to seven cubic feet of gasoline vapor per 100 cubic feet of air represents a flammable condition. Any spark or flame can touch off gasoline as a fire or explosion.

Gasoline should not be stored inside the home — it is too dangerous. It should be kept in a garage, toolhouse, or some other outside building. Storage in the basement is dangerous as a gas can could be tipped over. If the seal was imperfect, gasoline would leak out and begin to vaporize. Given certain circumstances, the whole basement could blow up — triggered by the flame from a furnace or water heater or from the arc of an electric switch.

Many states have recognized the hazards of gasoline around the home and are developing safety regulations that help to protect the public. The proper use of storage containers is one important area. Ideal storage is in a red metal can clearly labeled GASOLINE. Red is a universal color indicating danger. A metal can, such as shown in figure 2-1, will not break if it is accidentally dropped or struck. Follow these rules for the storage of gasoline:

- Store gasoline in a metal container.
- Store gasoline in an outbuilding — not inside the home.
- Do not store gasoline during the off season. It is too dangerous to have around.
- Do not allow small children to have access to the gasoline supply.
- Store gasoline away from flames, excessive heat, sparks caused by static electricity, sparks caused by electrical contacts, or sparks caused by mechanical contact.
- Have the proper portable fire extinguisher available for use if needed.

Fig. 2-1 Store gasoline and other flammable liquids in their proper containers.

Gasoline should be regarded as dangerous; using it safely involves decisions and judgments. Furthermore, gasoline should be regarded as fuel for gasoline engines and not as a handy all-purpose solvent, cleaner, or fire starter. Again, the volatility and explosive nature of gasoline make it unsuitable for any type of general household use. Gasoline used as a cleaning agent for paint brushes or to cut grease creates real safety hazards. Accidental ignition can occur while the gasoline is being used. Also, disposing of the dirty gasoline may be hazardous.

Never use gasoline to start a fire. The result may be a fire or explosion much larger than the intended size resulting in injury to the person with the gasoline.

Observe these precautions when using gasoline:

- Always regard gasoline as dangerous.
- Use gasoline only as a fuel for engines or devices where its use is clearly intended.
- Never smoke cigarettes, pipes, or cigars around gasoline.
- Never fill the tank of a hot engine or a running engine if the tank and engine are at all close to each other. Gasoline splashed on a hot engine can ignite or explode.

• When pouring gasoline from a container to the tank, reduce the possibility of a static electricity spark by having metal-to-metal contact. A metal spout held against the tank opening is good to use. A funnel that contacts both the container and the tank is also safe. Pouring gasoline through a chamois may present a static electricity hazard.

Fig. 2-2 Have fire extinguisher readily available and know the correct use of the various types.

FIRES

An understanding of fires and fire extinguishers is very useful, figure 2-2. Most fires fall into one of three categories:

Class A — Wood, cloth, paper, rubbish

Class B — Oil, gasoline, grease, paint

Class C — Electrical equipment

Class A Fires

Class A fires are the most common type of fire. Extinguishing Class A fires consists of quenching the burning material and reducing the temperature below that of combustion. Some extinguishing methods also have a smothering effect on the fire.

The Class A fire can be effectively fought with water from a fire hose, garden hose, or a water-base extinguisher. The water stream should be directed at the base of the fire first, then back and forth following the flames upward.

The dry chemical extinguisher is the most common extinguisher used in the home, office, or shop. Dry chemical extinguishers are effective in fighting almost all types of fires (Class A, B, or C). These extinguishers expel a chemical that interrupts and smothers the fire.

Class B Fires

Class B fires involve burning liquids such as gasoline, oil, paint, thinners, solvents, etc. Careless or unknowing action against Class B fires can place the fire fighter in great danger. Burning liquids such as gasoline are unpredictable; they can explode into a ball of fire as well as burn. The basic extinguishing technique is to cut off the oxygen supply which feeds the burning liquid and interrupt the flame without splashing the burning liquid around. Dry chemical extinguishers are suitable for fighting Class B fires.

Carbon dioxide (CO_2) extinguishers contain carbon dioxide under high pressure; they also are used to extinguish Class B fires. Carbon dioxide does not support combustion and, therefore, has the effect of smothering the flame. The extinguisher should be used with a slow sweeping action that travels from side to side, working to the back of the flame area. The discharge horn of the extinguisher becomes very cold during discharge and so should not be touched. Another hazard is that using the extinguisher in a small room may decrease the oxygen supply which is dangerous to the fire fighter.

Foam fire extinguishers discharge a water-base foam that can smother the fire. When using a foam extinguisher on Class B fires, direct the foam at the back of the fire allowing the foam to spread onto the flame area. This minimizes the possibility of splashing the flaming liquid of an open container fire.

Class C Fires

Class C fires involve electrical equipment. Water-base extinguishing materials are not suitable for putting out Class C fires. Water on or around an electrical fire creates a shock hazard. If the electrical energy can be completely and positively shut off, the fire is no longer considered a Class C fire; the method used to extinguish the fire then depends on the material that is burning. For example, shorted wires in an electrical outlet may be arcing and causing insulation to smolder. If the fire has spread to the wooden walls and if the electric current is turned off, the burning walls are considered a Class A fire. A dry chemical extinguisher would be suitable for both the electrical fire and the wood fire. Using water would be hazardous unless the electricity had been shut off. If the electrical energy cannot be completely shut off, the fire must be fought with carbon dioxide extinguishers, dry chemical extinguishers, or vaporizing liquid extinguishers.

CARBON MONOXIDE SAFETY

The hazard of carbon monoxide is very real. Carbon monoxide (CO) is a poison that is colorless, odorless and tasteless. It is a killer that is responsible for more poisoning deaths than any other deadly poison.

Carbon monoxide is the result of incomplete combustion of solid, liquid, or gaseous fuels of a carbonaceous nature. At home the gas may be present due to an improperly adjusted hot water heater. In industry carbon monoxide can be found with kilns, oven stoves, foundries, smelters, mines, forges, and in the distillation of coal and wood, to list a few. However, it also exists in a much more common circumstance — the exhaust gases of all internal combustion engines from lawn mowers to automobiles. When an engine is in operation, carbon monoxide is produced. Knowledge about this poison and how to eliminate or minimize the danger is an important topic for everyone.

The action of this poison on the human body can be quite rapid. The hemoglobin of the blood has a great attraction for CO, three hundred times greater than that for oxygen. When the CO is combined with the hemoglobin, it has the effect of reducing the amount of hemoglobin available to carry oxygen to the body tissues. If large amounts of CO combine with the hemoglobin, the body becomes starved for oxygen and literally suffocates.

Ventilation is of prime importance in preventing carbon monoxide poisoning. Unburned gases and exhausts must be carried away as effectively as possible by using chimneys, ventilation systems, and exhaust systems. These systems must be efficient because even small amounts of the gas can cause a dulling of the senses, which indicates danger.

Exhaust gases and carbon monoxide can get inside an automobile in several ways: through a defective exhaust system — tail pipe, muffler, or manifold; or through rusted-out or defective floor panels. If there is excessive CO around the automobile due to faulty exhausting or a poorly tuned engine, it can come in through an open window. For example, an open tailgate window on a station

Fig. 2-3 Ventilation is of prime importance in preventing carbon monoxide poisoning.

wagon can produce a circulation pattern that brings exhaust into the auto through the tailgate window. In the passenger section the carbon monoxide can build up to dangerous proportions. Station wagon rear windows should be closed, especially if other windows are open.

Consider the front window vents of the auto. When these are opened only slightly they draw air from the passenger compartment. If there is a hole at some exhaust concentration outside the compartment, exhaust can be drawn in. Generally, it is best to have the front vent windows open plus the side windows rolled down a bit in order to provide an adequate flow of fresh air. The air ducts that bring air in through the heater system are also a good ventilation source.

Knowing the symptoms of carbon monoxide poisoning and heeding their warnings could be a life-and-death matter. These symptoms include a tightness across the forehead followed by throbbing temples, weariness, weak-

ness, headache, dizziness, nausea, decrease in muscle control, increased pulse rate, and increased rate of respiration. If anyone has these symptoms while riding in a car, the car should be stopped and the person should get some fresh air.

If a person is discovered who has passed out from carbon monoxide poisoning, remove the victim to the fresh air immediately. If breathing has stopped or if the victim is only gasping occasionally, begin artificial respiration at once. Also, have someone call a doctor and/or emergency squad.

To prevent carbon monoxide poisoning, follow these rules:

- **Do not drive an auto with all the windows closed.**
- **Do not operate internal combustion engines in closed spaces such as garages or small rooms, unless the space is equipped with an exhaust system.**
- **Keep carbon monoxide producing engines and devices tuned or adjusted properly in order to reduce the output of CO gas.**
- **Do not sit in parked cars with the engine running.**
- **Remember the importance of properly working ventilation and exhaust systems.**

SAFETY WITH BASIC HAND TOOLS

Tools should be treated with respect and should be well cared for. They should be stored in a way that will protect their vital surfaces, cutting edges, and true surfaces. A wall tool panel with individual hangers or holding devices for each tool is ideal since the tools are easily seen, ready for use, and well protected. Tools carelessly thrown in a drawer can scratch, dull, or damage each other and are a hazard to the person who is looking for the right tool.

When tools must be carried in a toolbox, they should have individual compartments, if possible. If the box lacks individual compartments, the more sensitive tools and those with cutting edges should be wrapped in cloth.

Moisture and the rusting that results is a problem. If tools get wet, wipe them dry before storing. Also, the possibility of rusting can be lessened by wiping the tools with a slightly oily rag. If a tool picks up dirt during use, wipe it clean before putting it away.

The basic rules for the safe use of hand tools are:

> • **Use tools that are in good condition and well sharpened. A dull tool is inefficient and is more likely to slip during use.**
>
> • **Use the correct tool for the job and use it as it should be used. The incorrect use of tools causes many accidents.**

In some operations, hand tools can present the hazard of flying metal chips. A misstruck nail may ricochet across the room. Even a hammer can be dangerous. Especially dangerous are chipping operations with a cold chisel. All struck tools such as chisels and punches can mushroom out at the end over a period of use, figure 2-4. The ends should be dressed on a grinder to eliminate the possibility

BEFORE DRESSING   AFTER DRESSING

Fig. 2-4 All struck tools should be properly dressed on the end.

WRONG   RIGHT

Fig. 2-5 The hammer face should strike the work with its full, flat face.

of tool chips breaking and flying from the tool when it is hit. Hammers can chip if they are struck side blows or glancing blows; hammers should strike the work with their full face, figure 2-5.

When properly used, a wrench is a safe tool. However, if used incorrectly, the wrench can be dangerous. Be certain to use the exact size wrench for the nut or bolt to prevent slipping or damage to the flats of the nut or bolt. If an adjustable wrench is used, place it tightly against the flats of the nut or bolt and be sure that the direction of force places the major strain on the fixed jaw and not on the movable jaw, figure 2-6. Be sure to pull the wrench toward you; never push it. Pushing can be dangerous; the nut or bolt may suddenly loosen, or the tool may slip causing the user to be injured.

Screwdrivers are probably the most misused of all common tools. The list of abuses ranges from using it to open paint cans to using it as a wrenching bar. The tool is intended to be used for tightening or loosening various types of screws; its tip must be preserved for this purpose. Use a screwdriver that is the correct size, figure 2-7. It should fit the screw slot snugly in width; the blade should be as wide as the slot is long. Have several sizes of screwdrivers in a tool assortment. It is a poor practice to hold the work in one hand in such a manner that a slip would send the screwdriver into the palm of the hand. Screwdrivers, pliers, or other tools

WRONG

RIGHT

Fig. 2-6 Use an adjustable wrench so the strain is on the fixed jaw.

TOO SMALL

CORRECT SIZE

Fig. 2-7 Use the correct size screwdriver for the job.

that are used for electrical or electronic work should have insulated handles, figure 2-8.

Files are rarely sold with handles attached but they should be so equipped. The tang is fairly sharp and can cut or puncture the hand if the file suddenly hits an obstruction. The brittle nature of files makes them a poor risk and dangerous as a pry bar; do not use them for this purpose.

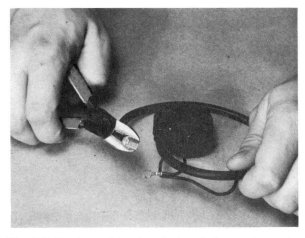

Fig. 2-8 Tools used in electrical and electronic work should have insulated handles.

SAFETY WITH BASIC MACHINES AND APPLIANCES

Safe use of basic household machines, mechanisms, and appliances during normal use, adjustment, and repair is important to all. Reference is made to power hand tools; power tools; appliances such as stoves, refrigerators, dishwashers, clothes driers, washing machines, air conditioners, electronic equipment such as television sets, record players, and radios; powered hobby or sports equipment; small home appliances for kitchen or personal use; and so forth. Several common rules should be followed when using these machines:

- Before attempting to adjust or repair a machine, always unplug or disconnect the machine or device from the electrical power source. Be certain that someone else does not accidentally reconnect or plug in the machine while you are working on it. If the power source is in a remote location, tag the source "Do not connect. Equipment under repair."

- Before attempting to adjust or repair machines driven by engines, stop the engine, disengage the clutch, and remove the spark plug lead to eliminate the possibility of accidentally restarting the engine.

- Be sure to ground electrical equipment that requires grounding. Use the three-prong plug on portable tools or devices and properly ground the convenience outlet.

- Keep loose clothing away from rotating parts that could grab the cloth and thereby pull the operator into the machine.

Fig. 2-9 This man is practicing proper safety techniques while using a weed eater. Note the heavy work clothes, rugged shoes, safety glasses and hardhat.

- Protect eyes with safety glasses if there is any chance of flying chips or breakage.

- Thoroughly guard all belts, pulleys, chains, gears, etc. Do not remove these guards or allow them to be removed by others.

- Study the instructions that come with the machine. Be certain that the operating principles and the safety precautions are understood before operating or repairing the machine.

- Use the machine only in the manner and for the purpose for which it was designed.

- During adjustment and repair, retighten all nuts, bolts, and screws securely.

- Unless qualified to be there, keep away from high voltage areas in electronics equipment.

SAFETY WITH SMALL GASOLINE ENGINES

Safety education about lawn mowers, chain saws, sports vehicles, outboard motors, and other small gas engines should begin at an early age. These machines are very powerful and if improperly used can easily injure a person.

Small engine applications are often a person's first exposure to gasoline engines and powered machinery. There is danger in the fuel and its use around the machine. There is also danger in the use of the machine itself. Besides learning how to start and stop an engine, an operator must also know general safety considerations. Trial and error is a very poor way to learn the use of any prime mover or machine.

Anyone who operates a small gasoline engine should understand and use these safety rules:

- **Do not make any adjustments or repairs to machinery being driven by an engine without first stopping the engine. Even with the engine stopped a slight turn over of the engine can accidently restart the engine, especially if it is warm. To prevent the accidental restarting of the engine when working on it, remove the spark plug cable and ground it. Also, remove the spark plug to prevent restarting the engine.**
- **Do not fill the gas tank when the engine is running or hot. If gasoline is spilled on a hot engine, a fire or explosion can result. If the job is not finished and the engine needs more gasoline, stop the engine and let it cool off before refilling the tank. It is a good idea to start each job with a full gas tank.**

- **Do not operate a gasoline engine in a closed building; the danger of carbon monoxide poisoning is very real.**
- **Read the equipment instruction booklet. Know and understand the equipment before operating it.**
- **Keep the equipment in perfect operating condition with all guards in place.**

Power Lawn Mower

The power lawn mower is an example of one small engine application which is widespread in its use. Because of the widespread use certain safety procedures should be followed when operating the mower.

The cutting blade on lawn mowers can be dangerous. The whirling blade of a rotary lawn mower can cut through large sticks and pick up and hurl foreign objects like projectiles. Studies show that about half of the accidents that occur with rotary power lawn mowers are caused by thrown objects. Most accidents that involve the operator result in injury to toes, fingers, and legs.

Persons who operate a power lawn mower accept a safety responsibility for themselves and for others. These safety rules should be followed:

- **When starting a lawn mower, be certain to stand clear with feet and hands away from the blade, figure 2-10.**
- **Do not mow wet grass as wet grass can be slippery. In the case of electric lawn mowers, wet grass can present an electrical shock hazard.**
- **Keep away from the grass discharge chute; sticks and stones are likely to be thrown out of this opening.**
- **When mowing hills or inclines, do not mow up and down; mow across the face of the slope.**

Fig. 2-10 Keep feet away from the blade when starting the mower.

- Do not pull a lawn mower — push it. If a person falls while pulling the mower, it may be pulled over the body.

- Inspect the lawn for rocks, sticks, wire and other foreign objects that can be converted into projectiles by the mower blade.

- Allow only a responsible person to operate a power mower.

- Small children should be kept away from the mowing area.

- Do not use a riding lawn mower as a play vehicle.

- Keep self-propelled lawn mowers under full control.

- Never leave a mower running unattended.

- Wear safety glasses while mowing.

REVIEW QUESTIONS

1. Name two factors that contribute to the safe operation of an engine.

2. What characteristics of gasoline makes it such a dangerous liquid?

3. List four rules for storing gasoline.

4. List three precautions to follow when using gasoline.

5. What type of extinguishers are best for putting out gasoline fires?

6. How is carbon monoxide produced? What is its effect on the body?

7. State two rules for the safe use of hand tools.

8. List six safety rules for operating power lawn mowers.

CLASS DISCUSSION TOPICS

- Discuss how safety becomes a part of a person's personality.
- Discuss how an individual's safety attitudes can affect other persons.
- Discuss why learning safety through trial and error is a poor practice.
- Discuss the proper use and storage of gasoline.
- Discuss the various ways to prevent carbon monoxide poisoning.
- Discuss equipment instruction booklets and their importance to safe machine operation.
- Discuss the dangers of loose clothing to mechanics, technicians, or machine operators.
- Discuss the proper care and storage of hand tools.

CLASS DEMONSTRATION TOPICS

- Demonstrate correct tool usage using several common hand tools.
- Demonstrate how worn or mushroomed tools can be repaired or dressed on a grinder.
- Demonstrate where to place hands and feet when starting a lawn mower.
- Illustrate proper attire for lawn mowing.
- Collect newspaper accounts of accidents to show how accidents happen when safety rules are violated.
- Demonstrate a fire's need for oxygen, by placing a glass over a lighted candle thus extinguishing the flame.

UNIT 3

Construction of the Small Gasoline Engine

OBJECTIVES After completing this unit, the student should be able to:
- Identify the basic parts of an engine and explain how they work together.
- Explain the principles of the four-stroke cycle.
- Explain the principles of the two-stroke cycle.
- Explain the differences in the construction of four-stroke cycle and two-stroke cycle engines.

The internal combustion engine is classified as a *heat* engine; its power is produced by burning a fuel. The energy stored in the fuel is released when it is burned. *Internal combustion* means that the fuel is burned inside the engine itself. The most common fuel is gasoline. If gasoline is to burn inside the engine, there must be oxygen present to support the combustion. Therefore, the fuel needs to be a mixture of gasoline and air. When ignited, a fuel mixture of gasoline and air burns rapidly; it almost explodes. The engine is designed to harness this energy.

The engine contains a cylindrical area commonly called the *cylinder* that is open at both ends. The top of the cylinder is covered with a tightly bolted-down plate called the *cylinder head.* The cylinder contains a *piston,* which is a cylindrical part that fits the cylinder exactly. The piston is free to slide up and down within the cylinder. The fuel mixture is brought into the cylinder, then the piston moves up and compresses the fuel into a small space called the combustion chamber. The *combustion chamber* is the area where the fuel is burned; it usually consists of a cavity in the cylinder head and perhaps the uppermost part of the cylinder. When the fuel is ignited and burns, tremendous pressure builds up. This pressure forces the piston back down the cylinder; thus the untamed energy of combustion is harnessed to become useful mechanical energy. The basic motion within the engine is that of the piston sliding up and down the cylinder, a *reciprocating* motion.

There are still many problems, however. How can the up-and-down motion of the piston be converted into useful rotary motion? How can exhaust gases be removed? How can new fuel mixture be brought into the combustion chamber? Studying the engine's basic parts can help answer these questions.

BASIC ENGINE PARTS

The internal combustion engine parts discussed here are those of the four-stroke cycle gasoline engine, the most common type. The most essential of these parts are briefly listed.

- *Cylinder:* hollow; stationary; piston moves up and down within the cylinder.
- *Piston:* fits snugly into the cylinder but still can be moved up and down.
- *Connecting rod:* connects the piston and the crankshaft.
- *Crankshaft:* converts reciprocating motion into more useful rotary motion.
- *Crankcase:* the body of the engine; it contains most of the engine's moving parts.
- *Valves:* "doors" for admitting fuel mixture and releasing exhaust.

Figure 3-1 shows the basic parts of two-cycle and four-cycle engines. Cross sections of two small gas engines are shown in figures 3-2 and 3-3.

Cylinder

The cylinder is a finely machined part which can be thought of as the upper part of the crankcase. The piston slides up and down in the cylinder. The fit is close; the piston is only a few thousandths of an inch smaller than the cylinder.

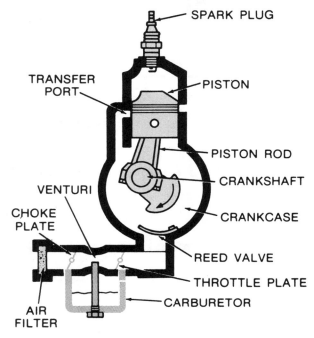

TWO-CYCLE

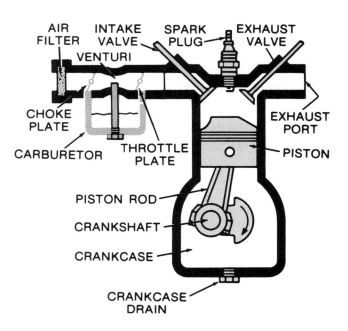

FOUR-CYCLE

Fig. 3-1 Basic parts of two-cycle and four-cycle engines

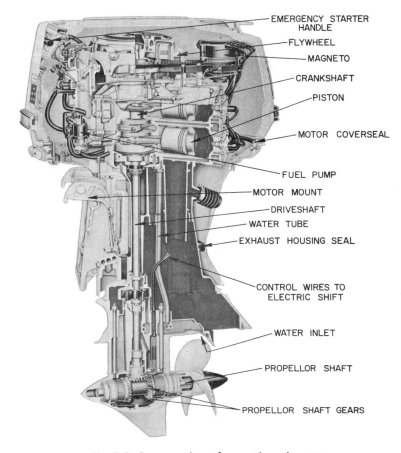

Fig. 3-2 Cross section of an outboard motor

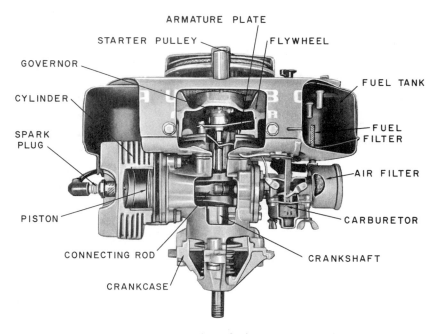

Fig. 3-3 Cross section of a lawn mower engine

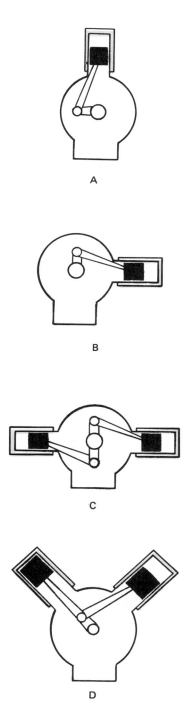

Fig. 3-4 Cylinder arrangement: (A) vertical, (B) horizontal, (C) opposed, and (D) V-type

Light duty aluminum engines often use the aluminum engine block for the cylinder. However, some aluminum engines have a cast iron cylinder sleeve cast into the engine and the two are inseparable. A cast iron engine has a cast iron block for the cylinder. Some engines have a chrome-plated cylinder.

In engine specifications, the diameter or width across the top of the cylinder is referred to as the *bore* of the engine; this is an important measurement. Figure 3-4 shows four different cylinder arrangements.

Cylinder Head

The cylinder head forms the top of the combustion chamber and is exposed to great heat and pressure. This is where the action is — the few cubic inches made up by the top of the piston, the upper cylinder walls and the cylinder head.

The cylinder head must be tightly bolted to the cylinder. It is important to tighten all cylinder head bolts with an even pressure and in their correct order so that uneven stress is not set up in the cylinder walls, figure 3-5. Such stress can distort or warp the cylinder walls.

Cylinder Head Gasket

A gasket between the two metal surfaces makes an airtight seal, figure 3-6. Gaskets are made of relatively soft material such as fiber, soft metal, cork, and rubber. Cylinder head gaskets must be able to hold the high pressure of combustion without blowing out and they must be heat resistant. The cylinder head also contains the threaded hole for the spark plug.

There are many engines that have the cylinder and the cylinder head cast as one piece, therefore, there are no gaskets or head bolts. A threaded spark plug hole is still located at the top of the cylinder head. Many small two-cycle engines use this design.

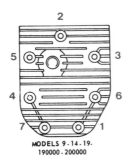

MODELS 9-14-19-
190000-200000

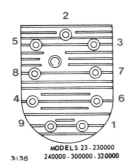

3138

MODELS 23-230000
240000-300000-320000

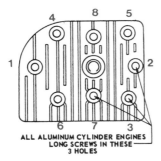

ALL ALUMINUM CYLINDER ENGINES
LONG SCREWS IN THESE
3 HOLES

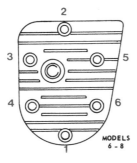

MODELS
6-8

Fig. 3-5 Cylinder head. Note the sequence for tightening the head bolts.

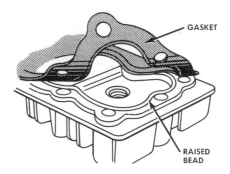

Fig. 3-6 The cylinder head gasket provides an airtight seal.

Piston

The piston slides up and down in the cylinder. It is the only part of the combustion chamber that can move when the pressure of the rapidly burning fuel mixture is applied. The piston can be made of cast iron, steel, or aluminum; aluminum is commonly used for small engines because of its light weight and its ability to conduct heat away rapidly.

There must be clearance between the cylinder wall and the piston to prevent excessive wear. The piston is 0.003 to 0.004 of an inch smaller than the cylinder. To check this clearance, a special feeler gauge is necessary.

The piston top, or crown, may be flat, convex, concave, or many other shapes. Manufacturers select the shape that causes turbulence of the fuel mixture in the combustion chamber and promotes smooth burning of the fuel mixture. The piston has machined grooves near the top to accommodate the piston rings, figure 3-7. To provide for expansion, some pistons are *cam ground* (not perfectly round).

The distance the piston moves up and down is called the *stroke* of the engine. The uppermost point of the piston's travel is called its *top dead center* (TDC) position. The lowest point of the piston's travel is called its

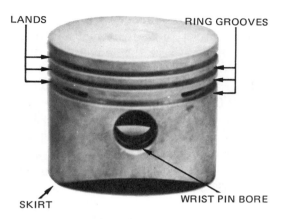

LANDS   RING GROOVES

SKIRT   WRIST PIN BORE

Fig. 3-7 The parts of a piston

bottom dead center (BDC). Therefore, stroke is the distance between TDC and BDC.

Piston Rings

Piston rings provide a tight seal between the piston and the cylinder wall. Without the piston rings much of the force of combustion would escape between the piston and cylinder into the crankcase. The piston rings mainly act as a power seal, ensuring that a maximum amount of the combustion pressure is used in forcing the piston down the cylinder. Since the piston rings are between the piston and cylinder wall, there is a small area of metal sliding against the cylinder wall. Therefore, the piston rings reduce friction and the accompanying heat and wear that are caused by friction. Another function of the piston ring is to control the lubrication of the cylinder wall.

The job of the piston rings might appear simple. However, the piston rings may have to work under harsh conditions such as (1) distorted cylinder walls due to improper tightening of head bolts, (2) cylinder out of round, (3) worn or scored cylinder walls, (4) worn piston, and (5) conditions of expansion due to intense heat. The important job of

providing a power seal can become difficult if the engine is worn or has been abused.

Piston rings are made of cast iron, steel or chrome and are finely machined. Sometimes the surfaces of piston rings are plated with other metals to improve their action. There are many designs of piston rings. Some rings, especially oil rings, may consist of more than one piece.

Blow-by and oil pumping are two serious problems that can be caused by bad piston rings, a bad piston, a warped cylinder, distortion, and/or scoring. In *blow-by,* the pressure of combustion is great enough to break the oil seal provided by the piston rings. When this happens, the gases of combustion force their way into the crankcase. The result is a loss in engine power and contamination of the crankcase.

Piston rings that are not fitted properly may start pumping oil. This pumping action leads to fouling in the combustion chamber, excessive oil consumption, and poor combustion characteristics. The rings have a side clearance which can cause a pumping action as the piston moves up and down.

A piston usually has three or four piston rings. Some have only one ring. The top piston ring is a compression ring, figure 3-8. This ring exerts a pressure of eight to twelve pounds on the cylinder wall. The ring has a clearance in its groove: it can move slightly up and down and expand slightly in and

Fig. 3-8 Compression ring

Fig. 3-9 Oil control ring

out. The second ring from the top is also a compression ring. These two rings form the power seal.

The third ring down (and fourth if one is present), is an oil control ring, figure 3-9. The job of this piston ring is to control the lubrication of the cylinder wall. These rings spread the correct amount of oil on the cylinder wall, scraping the excess from the wall and returning it to the crankcase. The rings are slotted and grooves are cut into the piston behind the ring. This enables much of the excess oil to be scraped through the piston and it then drips back into the crankcase. It should be noted that compression rings have a secondary function of oil control.

Two clearances (or tolerances) that are important in piston rings are end gap and side clearance. The end gap is the space between the ends of the piston ring, measured when the ring is in the cylinder. The gap must be large enough to allow for expansion due to heat but not so large that power loss due to blow-by will result. Often 0.004 inch is allowed for each inch of piston diameter (top ring), and 0.003 inch is allowed for rings under the top ring. Since the second and third rings are exposed to less heat their allowance can be smaller.

Side clearance or ring groove clearance is also provided for heat expansion, figure 3-10. Often 0.0025 inch is allowed for the top ring and 0.003 inch is allowed on the second and third rings.

Piston Pin or Wrist Pin

The piston pin is a precision ground steel pin that connects the piston and the connect-

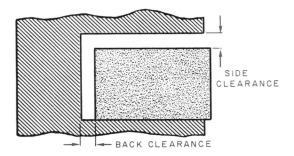

Fig. 3-10 Cross section showing a piston ring in its groove. Notice back clearance and side clearance.

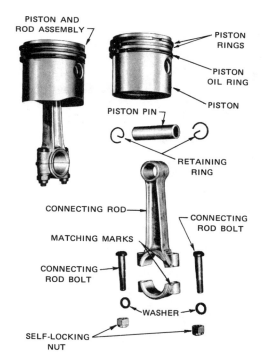

Fig. 3-11 Piston, piston pin, connecting rod, and associated parts

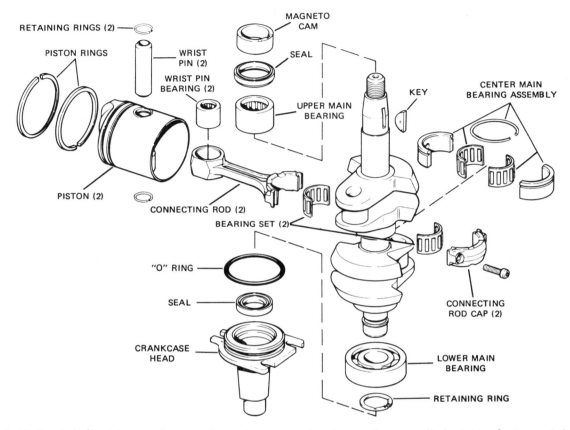

Fig. 3-12 Crankshaft, pistons, and connecting rod components — two-cycle, two-cylinder engine (note crankshaft)

ing rod, figure 3-11. This pin can be solid or hollow; the hollow piston pin has the advantage of being lighter in weight. Because piston pins are subjected to heavy shocks when combustion takes place, high-tensile strength steel is used. To keep noise at a minimum, the piston pin fits to a very close tolerance in the connecting rod. The piston pin does not rotate. It has a rocking motion similar to the action of the wrist when the wrist is moved back and forth, hence the name *wrist pin*. Heavy shocks, close tolerances, and the rocking motion make the lubrication of this part difficult. In most engines, lubricating oil is squirted or splashed on it.

Connecting Rod

The connecting rod connects the piston (with the aid of the piston pin) and the crankshaft, figure 3-12. Many small engines have cast aluminum connecting rods; larger engines can have steel connecting rods. Generally, the cross section of the connecting rod is an I-beam shape. The lower end of the connecting rod is fitted with a cap. Both the lower end and the cap are accurately machined to form a perfect circle. This cap is bolted to the rod, encircling the connecting rod bearing surface on the crankshaft. Many connecting rods are fitted with replaceable bearing surfaces called rod bearings or inserts; these are used especially on heavy-duty or more expensive engines.

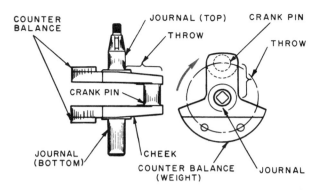

Fig. 3-13 Single-throw crankshaft: one crank pin for one cylinder

Crankshaft

The crankshaft is the vital part that converts the reciprocating motion of the piston into rotary motion. One end of the crankshaft has a provision for power takeoff (connecting accessory to use engine's power) and the other is machined to accept the flywheel.

Crankshafts are made of high-quality steel; they are carefully forged and then machined to close tolerances. Their bearing surfaces are large because they must accept a great amount of force. Since they revolve at high speeds, they must be well balanced. Figure 3-13 is an example of a single-throw crankshaft.

Figure 3-14 shows two types of small engine designs. Engines designed with *vertical* crankshafts are shown in figure 3-15. Vertical crankshafts are excellent for applications such as rotary power lawn mowers where the blade is bolted directly to the end of the crankshaft. Most wheeled applications such as motorbikes and go-carts use *horizontal* crankshaft engines which enable the power to be delivered to an axle through belts or chains. Multiposition engines such as chain saws are another variation. The crankshafts themselves are much the same but other changes in engine design are necessary to meet the requirements of crankshaft position. An engine can have

Fig. 3-14 Horizontal and vertical shaft engines

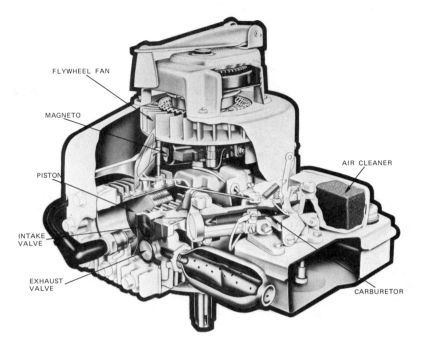

Fig. 3-15 Four-cycle engine with vertical shaft

FLYWHEEL FAN

MAGNETO

PISTON

INTAKE VALVE

EXHAUST VALVE

AIR CLEANER

CARBURETOR

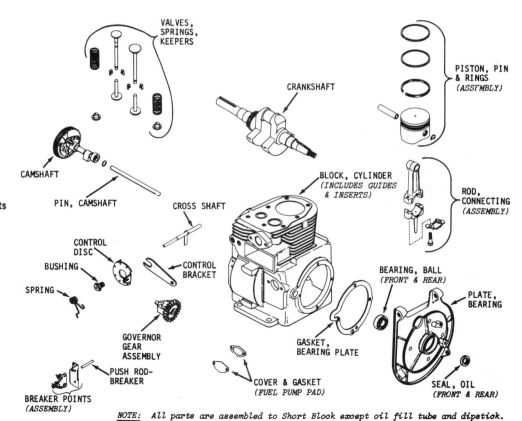

Fig. 3-16 Components of a short block assembly

VALVES, SPRINGS, KEEPERS

CRANKSHAFT

PISTON, PIN & RINGS *(ASSFMBLY)*

CAMSHAFT

PIN, CAMSHAFT

CROSS SHAFT

BLOCK, CYLINDER *(INCLUDES GUIDES & INSERTS)*

ROD, CONNECTING *(ASSEMBLY)*

CONTROL DISC

BUSHING

CONTROL BRACKET

SPRING

BEARING, BALL *(FRONT & REAR)*

PLATE, BEARING

GOVERNOR GEAR ASSEMBLY

GASKET, BEARING PLATE

PUSH ROD-BREAKER

BREAKER POINTS *(ASSEMBLY)*

COVER & GASKET *(FUEL PUMP PAD)*

SEAL, OIL *(FRONT & REAR)*

NOTE: All parts are assembled to Short Block except oil fill tube and dipstick.

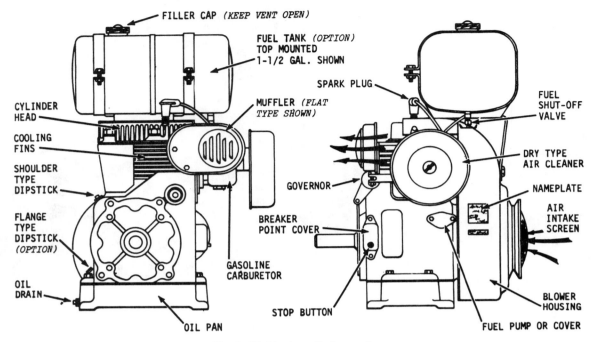

FILLER CAP *(KEEP VENT OPEN)*

FUEL TANK *(OPTION)*
TOP MOUNTED
1-1/2 GAL. SHOWN

SPARK PLUG

FUEL
SHUT-OFF
VALVE

CYLINDER
HEAD

MUFFLER *(FLAT
TYPE SHOWN)*

COOLING
FINS

DRY TYPE
AIR CLEANER

SHOULDER
TYPE
DIPSTICK

GOVERNOR

NAMEPLATE

AIR
INTAKE
SCREEN

FLANGE
TYPE
DIPSTICK
(OPTION)

BREAKER
POINT COVER

OIL
DRAIN

GASOLINE
CARBURETOR

STOP BUTTON

BLOWER
HOUSING

FUEL PUMP OR COVER

OIL PAN

Fig. 3-17 Single-cylinder engine

clockwise or counterclockwise crankshaft rotation, depending upon its application and design.

Crankcase

The crankcase is the body of the engine. It houses the crankshaft and has bearing surfaces on which the crankshaft revolves. The connecting rod, cam gear, camshaft, and lubrication mechanism are also located within the crankcase. The cylinder and bottom of the piston are exposed to the crankcase. On four-cycle engines a reservoir of oil is found within the crankcase. This lubricating oil is splashed, pumped, or squirted onto all the moving parts located in the crankcase. Figure 3-16 shows the components of a short block assembly. A single-cylinder engine is illustrated in figure 3-17.

Generally the crankcase is made of cast aluminum for engines that are for light duty and especially where a lightweight engine is needed. Crankcases for heavy duty engines are most often cast iron.

Flywheel

The flywheel is mounted on the tapered end of the crankshaft. A *key* keeps the two parts solidly together and they revolve as one part, figure 3-18. (The slot that the key fits into is called the *keyway*.) The flywheel is

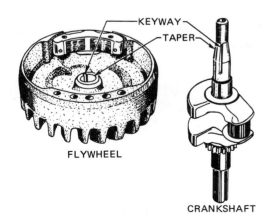

KEYWAY
TAPER

FLYWHEEL

CRANKSHAFT

Fig. 3-18 A key locks the flywheel and crankshaft together

found on all small engines. It is relatively heavy and helps to smooth out the operation of the engine. That is, after the force of combustion has pushed the piston down, the momentum of the flywheel helps to move the piston back up the cylinder. This tends to minimize or eliminate sudden jolts of power during combustion. The more cylinders the engine has, the less important this smoothing out action becomes.

The flywheel can also be a part of the engine's cooling system by having air vanes to scoop air as the flywheel revolves. This air is channeled across the hot engine by the cover or shroud, to carry away heat.

Most flywheels on small gasoline engines are part of the ignition systems; they have permanent magnets mounted in them. These permanent magnets are an essential part of the magneto ignition system.

Valves

Figure 3-19 shows the parts of a valve. The most common valves are called *poppet valves*. With these valves, exhaust gases can be re- moved from the combustion chamber and new fuel mixture can be brought in. To ac- complish this, the valve actually pops open and snaps closed. In the normal four-cycle combustion chamber there are two valves — one to let the exhaust out (exhaust valve) and one to allow the fuel mixture to come in (intake valve).

The two valves may look almost identical, but they are different. The exhaust valve must be designed to take an extreme amount of punishment and still function perfectly. Not only is it subjected to the normal heat of combustion (about 4500°F), but when it opens, the hot exhaust gases rush by it. When the valve returns to its seat, a small area touches, making it difficult for the valve's heat to be conducted away. This rugged valve opens and closes 1800 times a minute if the engine is operating at 3600 rpm. Exhaust valves are made from special heat-resisting alloys.

Valves can be made to rotate a bit each time they open and close. This action tends to prevent deposits from building up at the valve seat and margin. In addition, rotation provides for even wear and exposure to temperatures which greatly increases the life of the valves, figure 3-20.

Intake valves operate in the same manner as exhaust valves except they are not subjected

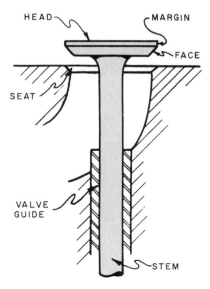

Fig. 3-19 A valve

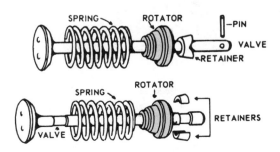

Fig. 3-20 "Rotocap" provides for even wear and long valve life

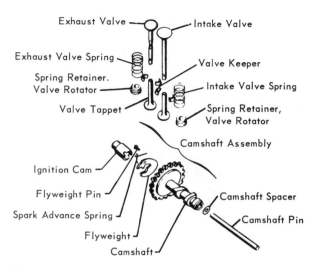

Exhaust Valve — Intake Valve

Exhaust Valve Spring — Valve Keeper

Spring Retainer. — Intake Valve Spring
Valve Rotator

Valve Tappet — Spring Retainer, Valve Rotator

Camshaft Assembly

Ignition Cam

Flyweight Pin

Spark Advance Spring — Camshaft Spacer

Flyweight — Camshaft Pin

Camshaft

Fig. 3-21 Complete valve train and camshaft assembly

Fig. 3-22 L-head design used for most four-stroke cycle small engines

to the extreme heat conditions. Each time they open new fuel mixture rushes by, helping to cool the intake valve.

Operation of the valve requires the help of associated parts — the valve train and camshaft, figure 3-21. The valve train consists of the valve (intake or exhaust), valve spring, tappet or valve lifter, and cam. The valve spring closes the valve, holding it tightly against its seat. The tappet rides on the cam; there is a small clearance between the tappet and the valve stem. Some tappet clearances are adjustable especially on larger and more expensive engines. The cam or lobe pushes up on the tappet; the tappet pushes up on the valve stem. The valve opens when the high spot on the cam is reached. The opening and closing of the valves must be carefully timed in order to get the exhaust out at the correct time and the new fuel mixture in at the correct time.

Valve Arrangement

Engines may be designed and built with one of several valve arrangements.

- *The L-head:* both valves open up on one side of the cylinder. This is the most common arrangement in small gasoline engines, figures 3-22 and 3-23.

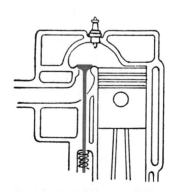

Fig. 3-23 L-head valve arrangement

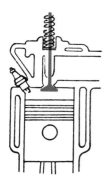

Fig. 3-24 I-head valve arrangement

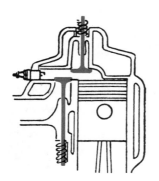

Fig. 3-25 F-head valve arrangement

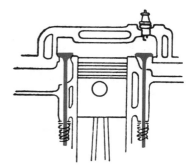

Fig. 3-26 T-head valve arrangement

- *The I-head:* both valves open down over the cylinder, figure 3-24.

- *The F-head:* one valve opens up and one valve opens down; both are on the same side of the cylinder, figure 3-25.

- *The T-head:* one valve is on one side of the cylinder and the other valve is on the opposite side; both valves open up, figure 3-26.

Camshaft

The camshaft and its associated parts control the opening and closing of the valves. The cams on this shaft are egg-shaped. As they revolve, their noses cause the valves to open and close. A tappet rides on each cam. When the nose of the cam comes under the tappet, the valve is pushed open admitting new fuel mixture or allowing exhaust to escape, depending on which valve is opened.

The camshaft is driven by the crankshaft. On four-cycle engines, the camshaft operates at one-half the crankshaft speed. (If the engine is operating at 3400 rpm, the camshaft revolves at 1700 rpm.) The crankshaft and camshaft must be in perfect synchronization. Therefore, most engines have timing marks on

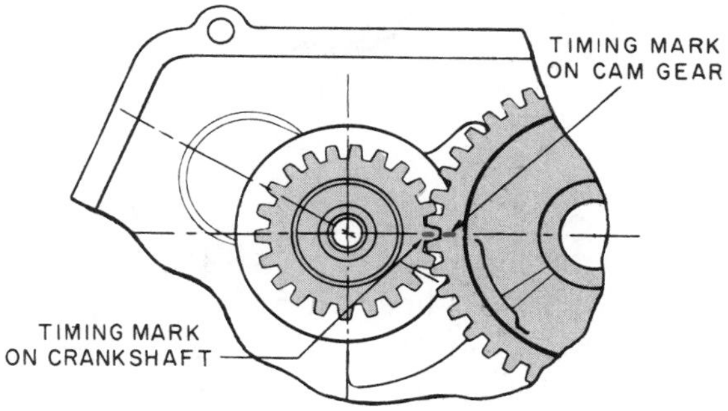

Fig. 3-27 Aligning timing marks on the crankshaft gear and camshaft gear. Notice the larger camshaft gear revolves at one-half crankshaft speed.

the gears to ensure the correct reassembly of the gears, figure 3-27.

FOUR-STROKE CYCLE OPERATION

All of these parts and many others must be assembled to operate the gasoline engine. The most common method of engine operation is the four-stroke cycle. Most small gasoline engines as well as automobile engines operate on this principle. The four-stroke cycle engine is a very successful producer of power.

Four-stroke cycle means that it takes four strokes of the piston to complete the operating

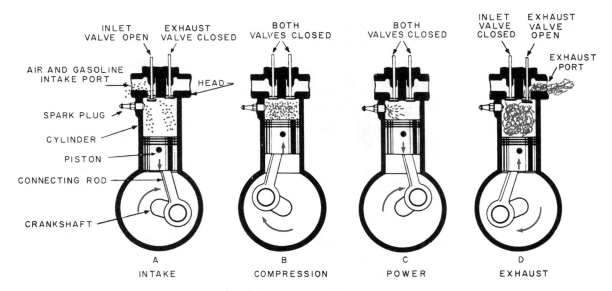

Fig. 3-28 The four-stroke cycle

cycle of the engine, figure 3-28. The piston goes down the cylinder, up, down, and back up again to complete the cycle. This takes two revolutions of the crankshaft; each stroke is one-half revolution.

Intake is when the piston travels down the cylinder and fuel mixture enters the combustion chamber. Reaching the bottom of its stroke, the piston comes up on compression and the fuel mixture is compressed into the combustion chamber. Power comes when the fuel mixture is ignited, pushing the piston back down. The piston moves up once more pushing the exhaust gases out of the combustion chamber. This is exhaust. These four strokes complete the engine's cycle and require two revolutions of the crankshaft. As soon as one cycle is complete, another begins. If an engine is operating at a speed of 3600 rpm there are 1800 cycles each minute.

Intake Stroke

If the fuel mixture is to enter the combustion chamber, the intake valve must be open and the exhaust valve must be closed. With the intake valve open and the piston traveling down the cylinder, the fuel mixture rushes in easily. This is the *intake stroke.*

Pushing down on everything is an atmospheric pressure of 14.7 pounds per square inch (psi) at sea level. When the piston is at the top of its stroke there is normal atmospheric pressure in the combustion chamber. When the piston travels down the cylinder there is more space for the same amount of air and the air pressure is reduced; a partial vacuum is created. The intake valve opens and the normal atmospheric pressure rushes in to equalize this lower air pressure in the combustion chamber. The fuel mixture is actually pushed into the combustion chamber, even though it seems to be sucked in by the partial vacuum.

Compression Stroke

When the piston reaches the bottom of its stroke and the cylinder is filled with fuel mixture, the intake valve closes and the exhaust

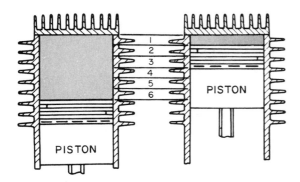

Fig. 3-29 An engine with a six-to-one compression ratio

valve remains closed. The piston now travels up the cylinder compressing the fuel mixture into a smaller and smaller space. When the fuel mixture is compressed into this small space it can be ignited more easily and it expands very rapidly.

The term *compression ratio* applies to this stroke. *Compression ratio* is the ratio of the total cylinder volume when the piston is at the bottom of its stroke to the volume remaining when the piston is at the top of its stroke. If the piston compresses the fuel mixture into one-sixth of the original space, the compression ratio is 6:1, figure 3-29. An engine whose piston compresses the fuel mixture into one-eighth the original space has a compression ratio of 8:1. The higher the compression, the more powerful the force of combustion.

Power Stroke

When the piston reaches the top of its stroke and the fuel mixture is compressed, a spark jumps across the spark plug igniting the fuel mixture. The fuel mixture burns very rapidly, almost exploding; the burning, expanding gases exert a great pressure in the combustion chamber. The combustion pressure is felt in all directions, but only the pis-

ton is free to react. The piston reacts by being pushed rapidly down the cylinder. This is the power stroke. *Power stroke* occurs when the rapidly burning fuel creates a pressure that pushes the piston down thus turning the crankshaft and producing usable rotary motion.

Both intake and exhaust valves must be closed for the power stroke. The force of combustion must not leak out the valves or blow-by the piston rings. Piston rings that fit properly prevent this. The cylinder head gasket and the spark plug gasket should be airtight. The power must be transmitted to the top of the piston and not lost elsewhere.

Exhaust Stroke

When the piston reaches the bottom of its power stroke, the momentum of the flywheel and crankshaft bring the piston back up the cylinder. This is the *exhaust stroke*. The exhaust valve opens, the intake valve remains closed, and the piston pushes the burned exhaust gases out of the cylinder and combustion chamber.

TWO-STROKE CYCLE OPERATION

Another common operating principle for gasoline engines is the two-stroke cycle engine, figure 3-30A. This engine is also very successful and in common use today, especially for outboard motors, chain saws, and many lawn mowers, figure 3-30B. The two-stroke cycle engine is constructed somewhat differently from the four-stroke cycle engine. See figure 3-31 for a comparison of moving parts.

Two-stroke cycle means that it takes two strokes of the piston to complete the operating cycle of the engine.

The basic two-stroke cycle consists of a pumping action in the crankcase — high and

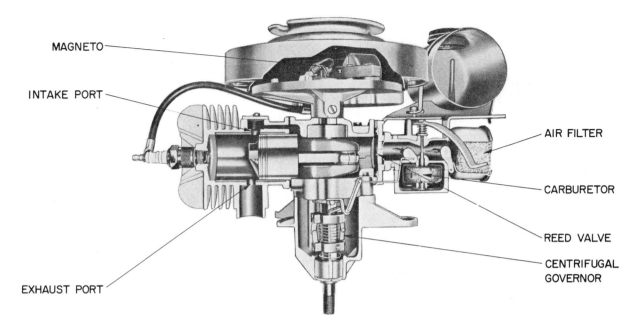

MAGNETO

INTAKE PORT

AIR FILTER

CARBURETOR

REED VALVE

CENTRIFUGAL GOVERNOR

EXHAUST PORT

Fig. 3-30A Two-cycle engine

Fig. 3-30B The lightweight and constant lubrication features of a two-cycle engine make it ideal for chain saws.

MOVING PARTS—4-CYCLE ENGINE

MOVING PARTS—2-CYCLE ENGINE

Fig. 3-31 Comparison of four-cycle and two-cycle engines

low pressure caused by the movement of the piston up and down the cylinder. On a four-stroke cycle engine the crankcase plays no part in fuel induction but on a two-stroke cycle engine the crankcase is a vital part.

The two-stroke cycle starts with the piston moving up the cylinder. As the piston moves up the cylinder, a low pressure or partial vacuum is created in the crankcase; there is a larger space for the same amount of air. The greater atmospheric pressure outside rushes through the carburetor pushing open the

springy reed valve, filling the crankcase with fuel mixture. When the piston is at the top of its stroke, the pressure in the crankcase and the atmospheric pressure are just about equal and the leaf valve springs shut.

As the piston moves back down the cylinder the fuel mixture trapped in the crankcase is put under pressure. The reed valve can only open in the opposite direction. Near the bottom of the piston's stroke the intake transfer ports are uncovered by the piston. The pressurized fuel mixture in the crankcase

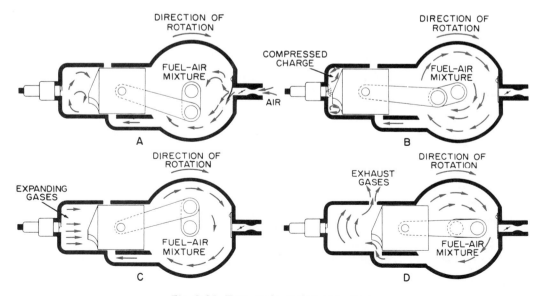

Fig. 3-32 Two-cycle engine operation

now has a path to escape into the cylinder and it rushes into the cylinder where it can be compressed and burned.

The fuel charge, now in the cylinder, is compressed as the piston moves up the cylinder. A spark ignites the fuel mixture and the great pressure of combustion pushes the piston rapidly down the cylinder. As the piston nears the bottom of its stroke the exhaust ports begin to uncover and the exhaust gases, which are still hot and under pressure, start to rush through the exhaust ports and out of the engine.

Just after the exhaust ports begin to uncover, the downward motion of the piston uncovers the intake transfer ports and compressed fuel mixture from the crankcase begins to rush in. Exhaust is leaving the cylinder at the same time new fuel mixture is entering. The incoming fuel mixture is directed toward the top of the cylinder and its action helps to scavenge out the exhaust gases. With the transfer of fuel complete, the piston moves up covering the intake transfer ports and the exhaust ports thus trapping the fuel mixture for compression.

The cycle is then repeated. If the engine operates at 4000 rpm, the cycle is completed 4000 times each minute. There is a power stroke for each revolution of the crankshaft.

To summarize the two-stroke cycle operation:

- Fuel mixture in the crankcase is put under pressure when the piston is going down on the power stroke.

- At the bottom of the power stroke, exhaust gases leave and fuel mixture is transferred from the crankcase to the cylinder.

- During the compression stroke, fuel mixture is brought into the crankcase.

Crankcase

The crankcase of a two-stroke cycle engine is designed to be as small in volume as possible. The smaller the volume, the greater the pressure created as the piston comes back down the cylinder. Crankcase pressure moves the fuel mixture through the engine. The greater the crankcase pressure, the more efficient the transfer of fuel from the crankcase through the transfer ports into the cylinder. The crankcase must be air tight. The only openings are the reed valves and the intake transfer ports which open only at specific times.

Moving parts in the crankcase must be lubricated; however, there is no reservoir of oil in the crankcase. Instead, the lubricating oil is mixed with the gasoline. A small amount of oil enters the crankcase each time the gasoline and air-fuel mixture enters the crankcase.

Intake Transfer Ports and Exhaust Ports

Intake transfer ports and exhaust ports are holes drilled in the cylinder wall which allow exhaust gases to escape and new fuel mixture to enter from the crankcase. The ports are located near the bottom of the cylinder and are covered and uncovered by the piston. By using ports, many parts seen in the four-cycle engine are eliminated: the valves, tappets, valve springs, cam gears, and camshaft. The remaining major moving parts are the piston, connecting rod, and crankshaft.

Piston (Cross Scavenged)

The piston still serves the same function in the cylinder, but for two-cycle engines the top is usually designed differently, figure 3-33. The two-stroke cycle pistons have a contoured top. On the intake side there is a sharp deflection that sends the incoming fuel mixture to the top of the cylinder. On the exhaust side there is a gentle slope so that exhaust gases have a clear path to escape. The loop scavenged two-cycle engine does, however, have a flat top piston.

Loop Scavenging

The loop scavenged two-cycle engine is basically the same as any other two-cycle engine. However, it does not need the contoured piston of the cross scavenged design because it does not have to deflect incoming gases. The difference is that the intake ports are drilled into the cylinder walls at an angle which aims the incoming fuel mixture toward the top of the cylinder. Usually there are two pairs of intake ports angling in from the sides directing the fuel mixture to the top of the cylinder around and back through the center of the cylinder to scavenge out the last of the exhaust through the exhaust ports. Loop scavenging provides a more complete removal of exhaust gases and produces somewhat more horsepower per unit weight.

Reed or Leaf Valves

Reed, or leaf valves, are located between the carburetor and crankcase. The valve itself is a thin sheet of springy alloy steel. The valve is pushed open by atmospheric pressure which allows fuel mixture to enter the crankcase; it springs closed to seal the crankcase. This occurs when a low pressure is created in the crankcase by the action of the piston moving up the cylinder. There may be only one reed valve or there may be several reed valves working together. The majority of small gasoline engines operating on the two-cycle principle use reed valves even though it is possible to use a poppet valve in the crankcase. Several kinds of reed plates are shown in figure 3-34.

Not all two-cycle engines use reed or leaf valves. Some engines have the compression and power strokes with intake and exhaust taking place between the two but use different valves to achieve the result.

Fig. 3-34 Reed plates

Fig. 3-33 A piston for a two-cycle engine showing the exhaust side and the intake side

Rotary Valve

Some two-cycle engines use a rotary valve for admitting fuel mixture, figure 3-35. The rotary valve is a flat disc with a section removed; it is fastened to the crankshaft. The rotary valve normally seals the crankcase. However, as the piston nears the top of its stroke and there is a slight vacuum in the crankcase, the valve is rotated to its open position. This allows fuel mixture to travel from the carburetor through the open valve and into the crankcase.

Third Port Design

Third port design has the regular exhaust ports and intake ports on the cylinder wall but in addition it has a third port in the cylinder wall, figure 3-36. This third port is for admitting fuel mixture to the crankcase. The bottom of the piston skirt uncovers the third port as the piston nears the top of its stroke. The piston's upward travel creates a low pressure in the crankcase. When the port

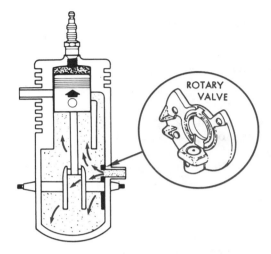

Fig. 3-35 Rotary valve

is uncovered, fuel mixture rushes into the crankcase. As the piston comes down on the power stroke, the port is sealed off and the trapped mixture is pressurized — ready for transfer through the intake ports into the cylinder.

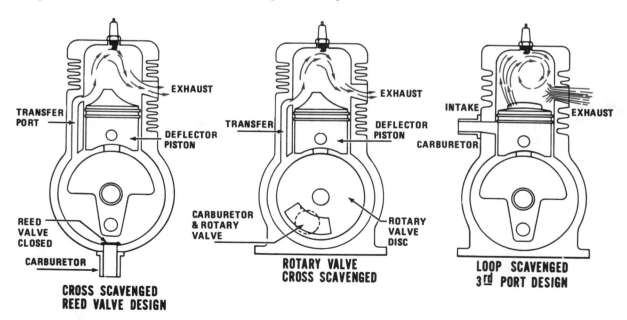

Fig. 3-36 Basic designs of two-cycle spark ignition engines

Poppet Valves

There are some two-cycle engines that use poppet valves to admit new fuel mixture into the crankcase. These poppet valves may be spring loaded and operated by differences in crankcase pressure. They open when the partial vacuum in the crankcase overcomes a slight spring tension. This happens when the piston is on its upward stroke. With the poppet valve open, fuel mixture rushes into the crankcase. The poppet valve may also be operated by cam action. In this case a crankshaft cam opens the valve for the piston's upward stroke.

Fig. 3-37 An all-terrain-vehicle has a powerful small gas engine.

REVIEW QUESTIONS

1. Explain the term *internal combustion.*
2. List the most essential moving parts of the four-stroke gasoline engine.
3. List the most essential stationary parts of the four-stroke gasoline engine.
4. Define the terms: bore, stroke, TDC, and BDC.
5. What purpose does a gasket serve?
6. What is a combustion chamber?
7. What two purposes do piston rings serve?
8. What is the function of the crankshaft?
9. What parts do the connecting rod and wrist pin work with?
10. What does the flywheel accomplish?
11. What does a key and keyway do?
12. Explain how the valves operate.
13. Explain what a tappet does.

14. Prepare a sketch of a cam.

15. On a four-cycle engine, at what speed does the camshaft revolve relative to the crankshaft?

Four-Stroke Cycle Review Questions

1. Explain what causes the fuel mixture to rush into the cylinder during the intake stroke.

2. What is accomplished on the compression stroke.

3. Explain the power stroke.

4. How many revolutions are required for the complete cycle?

Two-Stroke Cycle Review Questions

1. What parts are replaced by the intake and exhaust ports?

2. What are ports?

3. How are some two-stroke cycle pistons different from four-stroke cycle pistons?

4. Explain the action of the reed or leaf valves.

5. Explain how intake and exhaust take place on a two-cycle engine.

6. What part does the crankcase play in regard to the fuel mixture?

7. How many revolutions are required for the complete cycle?

8. What is the advantage of loop scavenging?

CLASS DISCUSSION TOPICS

• What other systems does an engine need in addition to the basic moving parts?

• Discuss the basic differences in two-cycle and four-cycle engines.

• Discuss the advantages and disadvantages of two-cycle and four-cycle engines.

• Discuss how parts can be damaged by careless handling.

CLASS DEMONSTRATION TOPICS

- Inspect the basic parts of the two-cycle and four-cycle engine.

- Remove the cylinder head of a four-cycle engine, rotate the crankshaft, and observe the action of the valves.

- Remove the valve spring cover of a four-cycle engine, rotate the crankshaft, and observe the action of the valve lifters and valve springs.

- Remove the carburetor of a two-cycle engine. Turn the engine over rapidly, and observe the action of the leaf valves. This action is not pronounced and must be carefully observed.

- Using a disassembled engine, connect the piston, connecting rod, and crankshaft to illustrate how reciprocating motion is converted to rotary motion.

- Using a camshaft and crankshaft, point out timing marks.

- Using a camshaft and crankshaft, count the gear teeth to illustrate the speed relationship of the two gears.

- Using a flywheel and crankshaft, point out the keyways.

- Using a disassembled engine, lay out the parts of the valve train.

- Trace the path of the fuel mixture through both two-cycle and four-cycle engines.

LABORATORY EXPERIENCE 3-1
ENGINE DISASSEMBLY (FOUR-STROKE CYCLE)

The gasoline engine consists of many parts that function together as a smooth power team. It is important to be able to identify these parts and know how they fit and work together. Every student should have an appreciation of the precision workmanship that goes into an engine. The integrity of the engine's precision should be preserved as the student disassembles and reassembles the parts.

STUDENT ASSIGNMENT

You are to disassemble a typical four-stroke cycle engine, study the basic parts and then reassemble the engine. Remember, speed in engine disassembly is *not* the important thing. What is important is a careful, workmanlike approach to the job. It is essential that you record your work in a work record box as you complete each step of the disassembly procedure.

Disassembly Procedure

Instructors may want to supplement or revise specific steps of this procedure since there are many makes of engines. The following disassembly procedure is a general guide.

1. Drain the oil from the crankcase.
2. Disconnect the spark plug cable and remove the spark plug.
3. Drain the gas tank, disconnect the fuel line, and remove the gas tank.
4. Remove the air cleaner. (Use care if it contains oil.)
5. Drain the carburetor, remove throttle and governor connections, and remove the carburetor from the engine.
6. Remove metal air shrouding.
7. Remove the flywheel nut; also remove grass screens, starter pulleys, etc.
8. Remove the flywheel.
9. Remove the magneto plate assembly. (The main bearing is on the flywheel side of this plate.)
10. Remove the cylinder head.
11. Remove the crankcase from the engine base. (This may not be necessary if the engine has an inspection plate.)

12. Unbolt the connecting rod cap and push the piston and rod up and out of the cylinder.

13. Remove the crankshaft by pulling it out. (The main bearing plate may have to be loosened to do this.)

WORK RECORD BOX

PART	DISASSEMBLY (nuts, bolts, etc.)	OPERATION PERFORMED	TOOL USED

Reassembly Procedure

Reverse the disassembly procedure. Listed are some points to remember.

1. Tighten all machine screws and bolts securely.

2. Be certain the piston is reinstalled exactly as it came out.

3. Line up the timing marks on the crankshaft with the timing mark on the camshaft.

4. Line up the match mark on the connecting rod cap with the match mark on the connecting rod.

5. Be certain the flywheel keyway slips into the key on the crankshaft.

6. Check all gaskets; replacement may be necessary, especially if the engine is to be operated.

7. Refill the crankcase with oil if the engine is to be operated.

8. Refill the gas tank if the engine is to be operated.

REVIEW QUESTIONS

Study the basic parts of the engine and answer the following questions.

1. Why should most gaskets be replaced when the engine is reassembled?

2. Explain the value of match marks on the connecting rod and cap.

3. Why is uniform tightening of cylinder head bolts important?

4. Explain why timing marks are used on the crankshaft and camshaft.

5. How many piston rings are on the piston? What type are they?

6. Why is it important to replace the key in the keyways?

7. Does the engine have a cast iron or an aluminum cylinder?

8. Were you certain the piston went back into the cylinder in the same position that it came out? How did you determine this?

9. Explain why cleanliness is important in engine work.

LABORATORY EXPERIENCE 3-2
VALVE TRAIN AND CAMSHAFT (FOUR-STROKE CYCLE)

The engine's exhaust and intake valves provide a route for exhaust gases to escape and a route for new fuel mixture to enter. The valves are of the poppet type, being raised off their seats when they are opened. They are subjected to severe punishment in the combustion chamber but still must function perfectly, opening at exactly the right instant and closing to make an airtight seal. The parts that work with the valves are referred to as the valve train and camshaft.

NOTE: This laboratory experience may be done at the completion of disassembly on Laboratory Experience 3-1, or it may be done starting with a fully assembled engine. If students start with a fully assembled engine they should follow the disassembly procedure for Laboratory Experience 3-1; then continue on with the disassembly of the valve train and camshaft.

STUDENT ASSIGNMENT

You are to remove the valves, the valve springs, and the valve spring retaining parts and then remove the camshaft and valve lifters. Study these parts and then reassemble them correctly. It is essential that you record your work in a work record box as you complete each step of the disassembly procedure.

Disassembly Procedure

Instructors may want to supplement or revise specific steps of this procedure since there are many makes of engines. The following disassembly procedure is a general guide.

1. Remove the valve plate exposing the valve springs, valve lifters, and valve stems.

2. Compress the valve spring with a valve spring compressor.

3. Remove the valve spring retainers by slipping or flipping them out, figure 3-38.

4. Pull the valve out of the engine.

5. Remove the valve spring, still compressed.

6. Remove the camshaft. If the camshaft is held in the crankcase with a camshaft support pin it must be driven out with a blunt punch. In most cases, drive the punch from the takeoff side toward the flywheel side.

   NOTE: Other engines may be constructed with a one-piece camshaft that comes out when the crankshaft is removed.

7. Pull out the camshaft.

8. Remove the valve lifters; they will probably fall out when the camshaft is removed.

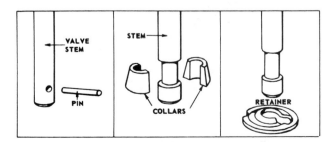

Fig. 3-38 Valve spring retainers

WORK RECORD BOX

PART	DISASSEMBLY (nuts, bolts, etc.)	OPERATION PERFORMED	TOOL USED

Reassembly Procedure

Reverse the disassembly procedure. Listed are some points to remember.

1. Reinstall the intake and exhaust valves in their correct places.
2. On some engines a special tool is needed to aid in the installation of valve spring retainers.
3. Reinstall the camshaft support pin correctly (from the flywheel side to the power takeoff side, in most cases).

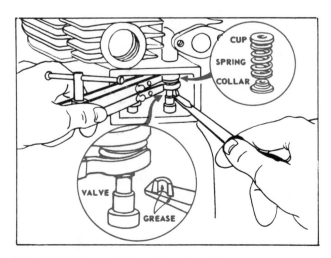

Fig. 3-39 One technique for reinstalling valve spring retainers

REVIEW QUESTIONS

Study the valves and associated parts and answer the following questions.

1. Why are exhaust valves and intake valves made differently?
2. Examine the two valves. How can exhaust valves be differentiated from intake valves?
3. What is the function of the valve spring?
4. Why does the camshaft rotate at one-half the crankshaft speed?
5. How many degrees apart are the intake and exhaust cams? Why?
6. What is the function of the tappet?
7. A four-cylinder engine has how many valves?

LABORATORY EXPERIENCE 3-3
EXHAUST AND INTAKE PORTS (TWO-STROKE CYCLE)

The exhaust ports and intake ports replace the valves of the four-stroke cycle engine. The ports are the exits for exhaust gases and new fuel mixture. The ports are actually holes bored in the cylinder wall. The action of the piston moving up and down the cylinder seals off and opens the ports.

STUDENT ASSIGNMENT

You are to remove the exhaust baffle plate, muffler, or exhaust manifold (different terms, all pertaining to the same area) and examine the ports. Observe how the ports are covered and uncovered by the piston when the engine is turned over. It is essential that you record your work in a work record box as you complete each step of the disassembly procedure.

Disassembly Procedure

Instructors may want to supplement or revise specific steps of this procedure since there are many makes of engines. The following disassembly procedure is a general guide.

1. Drain the gas tank.
2. Drain the carburetor.
3. Remove the spark plug cable from the spark plug.
4. Remove the spark plug from the cylinder.
5. Remove any metal shrouding that is around the exhaust area.
6. Remove the exhaust baffle plate, muffler, or exhaust manifold, thus exposing the exhaust ports.

WORK RECORD BOX

PART	DISASSEMBLY (nuts, bolts, etc.)	OPERATION PERFORMED	TOOL USED

Reassembly Procedure

Reverse the disassembly procedure.

REVIEW QUESTIONS

Study the exhaust ports and their operation and answer the following questions.

1. If the engine has been in operation, are there carbon deposits around the exhaust ports? If so, describe these deposits.
2. Explain why the engine turned over easily when the spark plug was removed.
3. How many ports make up the exhaust port team?
4. Is the exhaust side of the piston a gentle slope or a sharp deflection? Why?
5. Is loop scavenging used on the engine?

LABORATORY EXPERIENCE 3-4
REED VALVES (TWO-STROKE CYCLE)

The reed valves open to allow fuel mixture to rush into the crankcase and close to trap the fuel mixture in the crankcase. The part is made of a thin springy metal sheet that reacts to small differences in air pressure. It has no comparable part on the four-stroke cycle engine.

STUDENT ASSIGNMENT

You are to remove the reed valve plate, study the reed valves, and then reassemble the parts. It is essential that you record your work in a work record box as you complete each step of the disassembly procedure.

Observation of the opening and closing of the reed valves can usually be done by removing the carburetor, leaving the reed valve plate installed on the crankcase. Then, turn the engine over rapidly. The action of the reed valves is not pronounced, therefore, they must be watched very closely. If the spark plug is removed from the cylinder the engine can be turned over more easily.

Disassembly Procedure

Instructors may want to supplement or revise specific steps of this procedure since there are many makes of engines. The following disassembly procedure is a general guide.

1. Drain the gas tank.
2. Remove any metal shrouding from the carburetor area.
3. Drain the carburetor.
4. Remove throttle and governor connections.
5. Remove the carburetor.
6. Remove the spark plug from the cylinder.
7. Rotate the flywheel rapidly and observe the reed valve action.
8. Remove the reed valve plate from the crankcase.
9. Remove individual reed valves for inspection only if the engine is not to be operated again. The reed valves can be damaged through unnecessary handling.

WORK RECORD BOX

PART	DISASSEMBLY (nuts, bolts, etc.)	OPERATION PERFORMED	TOOL USED

Reassembly Procedure

Reverse the disassembly procedure.

REVIEW QUESTIONS

Study the reed valves and answer the following questions.

1. The reed valves are between what two main parts?

2. What condition in the crankcase causes the reed valves to be pulled open?

3. What condition in the crankcase causes the reed valves to spring closed?

4. How many individual reed valves work together as a team on the engine?

5. Trace the path of the fuel mixture in a two-cycle engine, beginning with the gasoline tank and continuing to the combustion chamber.

UNIT 4

Fuel Systems, Carburetion, and Governors

OBJECTIVES   After completing this unit, the student should be able to:

- Identify the basic parts of a fuel system and explain the function of each part.
- Identify the basic parts of the carburetor and explain the function of each.
- Discuss the theory of carburetor operation.
- Discuss several common types of carburetors.
- Discuss how gasoline vaporizes and burns and common problems such as vapor lock, detonation and preignition.
- Identify the basic parts of air vane and mechanical governors and explain how they work.
- Adjust the carburetor for maximum power and efficiency.
- Adjust the governed speed and discuss the action of the governor on the throttle.

The fuel system must maintain a constant supply of gasoline for the engine. The carburetor must correctly mix the gasoline and air together to form a combustible mixture which burns rapidly when ignited in the combustion chamber.

A typical fuel system contains a *gasoline tank* (reservoir for gasoline); a *carburetor* (mixing device for gasoline); the *fuel line* (tubes made of rubber or copper through which the gasoline passes from the gas tank to the carburetor); and an *air cleaner* (device for filtering air brought into the carburetor), figure 4-1. In addition, the system may have a *shutoff valve* (valve at the gas tank that can cut off the gasoline supply when the engine is not in use), figure 4-2; a *fuel pump* (pump that supplies the carburetor with a

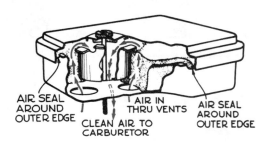

Fig. 4-1 Air cleaner

constant supply of gasoline); a *sediment bowl* (small glass bowl attached to the fuel line where dirt and other foreign matter can settle out), figure 4-3; and a *strainer* (fine screen in the gas tank to prevent leaves and dirt from entering the fuel line).

It should be noted that not all fuel systems contain all eight of the basic parts. Shut-off valves, sediment bowls, fuel pumps, and air cleaners are not common to all engines.

A constant supply of gasoline must be available at the carburetor. To provide this supply, four systems are in common use today. These systems are the suction system, gravity system, fuel pump system, and pressurized tank system.

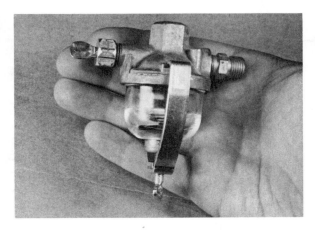

Fig. 4-3 Sediment bowl with built-in
fuel shut-off valve

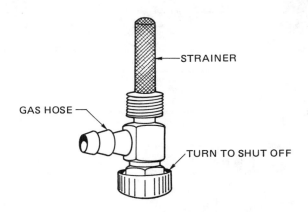

Fig. 4-2 Fuel shutoff valve

SUCTION SYSTEM

The suction system is the simplest fuel system, figure 4-4. With this system, the gas tank is located below the carburetor and the gasoline is sucked up into the carburetor. However, the gas tank cannot be very far away from the carburetor or the carburetor action will not be strong enough to pull the gasoline from the tank.

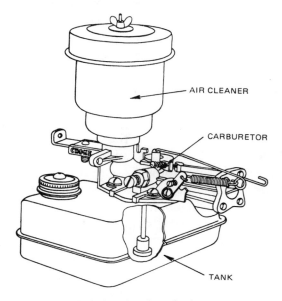

Fig. 4-4 Suction feed fuel system

GRAVITY SYSTEM

With the gravity system the gas tank is located above the carburetor and the gasoline runs downhill to the carburetor, figure 4-5. To prevent gasoline from continuously pouring through it, the carburetor has incorporated a float and float chamber. The *float chamber* provides a constant level of gasoline without flooding the carburetor. When gasoline is used, the float goes down, opening a valve to admit more gasoline to the float chamber. The float rises and shuts off the gasoline when it reaches its correct level. In actual practice the float and float valve do not rapidly open and close but assume a position allowing the correct amount of fuel to constantly enter the float chamber, figure 4-6. If the engine is speeded up, a new position is assumed by the float and float valve supplying an increased flow of gasoline.

FUEL PUMP SYSTEM

On many engines it is necessary to place the gas tank some distance from the carburetor. Therefore, the fuel must be brought to the carburetor by some means other than gravity or suction. One method is by using a fuel pump. The automobile engine uses a fuel pump. The outboard motor with a remote fuel tank is a good example of a smaller engine that commonly uses a fuel pump.

The fuel pump used on many two-stroke cycle outboard motors is quite simple. The fuel pump consists of a chamber, inlet and discharge valve, a rubber diaphragm, and a spring. It is operated by the crankcase pressure. As the piston goes up, low pressure in the crankcase pulls the diaphragm toward the crankcase sucking gasoline through the

Fig. 4-5 Gravity feed fuel system

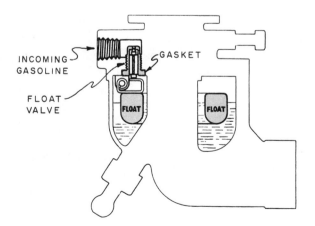

Fig. 4-6 Cutaway of a carburetor showing float and float valve

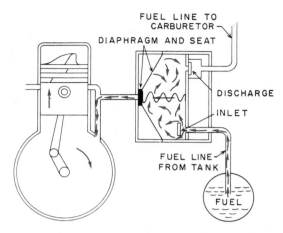

Fig. 4-7 Low pressure in the crankcase allows the fuel chamber to be filled.

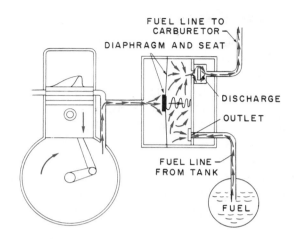

Fig. 4-8   High pressure in the crankcase forces the trapped fuel into the carburetor.

inlet valve into the fuel chamber, figure 4-7. As the piston comes down, pressure in the crankcase pushes the diaphragm away from the crankcase. When this happens, the intake valve closes and the discharge valve opens allowing the trapped fuel to be forced to the carburetor, figure 4-8.

PRESSURIZED FUEL SYSTEM

Another method for forcing fuel to travel long distances to the carburetor is the pressurized fuel system, figure 4-9. This method is sometimes used with outboard motors. Engines using this system have two hoses or lines between the gas tank and the engine. One line brings gasoline to the carburetor; the other line brings air under pressure from the crankcase to the gas tank. If this method is used, the gas tank must be airtight so that enough air pressure can build up to force the gasoline to flow to the carburetor.

THE CARBURETOR

The carburetor must prepare a mixture of gasoline and air in the correct proportions for

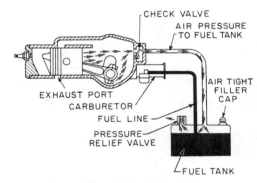

Fig. 4-9 Pressurized fuel systems can be found on some outboard engines.

burning in the combustion chamber, figures 4-10 and 4-11. The carburetor must function correctly under all engine speeds, under varying engine loads, in all weather conditions, and at all engine temperatures. To meet all these requirements, carburetors have many built-in parts and systems. Before studying the carburetor's operation in detail, an examination of the basic carburetor parts and their functions is helpful.

Throttle.   The throttle controls the speed of the engine by controlling the amount of fuel mixture that enters the combustion chamber. The more fuel mixture admitted, the faster the engine speed.

Choke. The choke controls the air flow into the carburetor. It is used only for starting the engine. In starting the engine, the operator closes the choke, cutting off most of the engine's air supply. This produces a *rich mixture* (one containing a higher percentage of gasoline) which ignites and burns readily in a cold engine. As soon as the engine starts, the choke is opened.

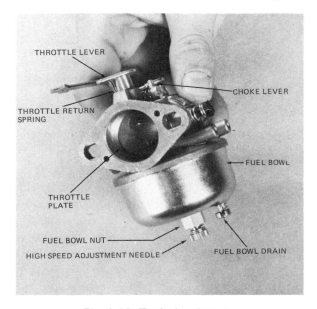

Fig. 4-10 Typical carburetor

Needle Valve. The needle valve controls the amount of gasoline that is allowed to pass out of the carburetor into the engine, figure 4-12. It controls the richness or leanness of the fuel mixture.

Idle Valve. The idle valve also controls the amount of gasoline that is available to the carburetor but it functions only at low speeds or idling. Some carburetors have what is called a *slow-speed needle valve* which performs basically the same job as the idle valve.

Float and Float Bowl. The float and float bowl are found on all carburetors except the suction-fed carburetor and diaphragm carburetors. The float and float bowl maintain a constant gasoline level in the carburetor.

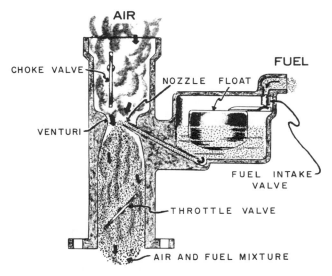

Fig. 4-11 Cross section of a simplified carburetor

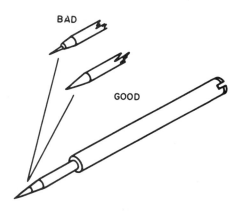

Fig. 4-12 Needle valves

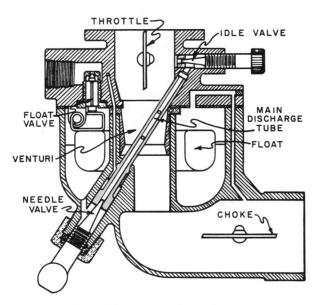

Fig. 4-13 A gravity fed carburetor

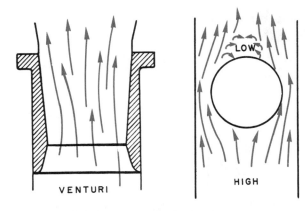

Fig. 4-14 (A) The venturi creates a low pressure in the carburetor. (B) The principle of the air foil also helps create a low pressure.

Venturi. The venturi is a section of the carburetor that is constricted or has a smaller cross-sectional area for the air to flow through. In the venturi, the gasoline and air are brought together and begin to mix.

Jets. Carburetor jets are small openings through which gasoline passes within the carburetor.

CARBURETOR OPERATION

In the beginning, air flows through the carburetor because there is a partial vacuum or suction in the combustion chamber as the piston travels down the cylinder. One usually thinks of the air being sucked into the engine by the piston action. Atmospheric pressure outside the engine actually pushes the air through the carburetor to equalize the lower pressure that is in the combustion chamber.

Air flows rapidly through the carburetor as the piston moves down. In the carburetor the air must pass through a constriction called the venturi, figure 4-13. For the same amount of incoming air to pass through this smaller opening, the air must travel faster and this it does. Here a principle of physics comes into use: the greater the velocity of air passing through an opening, the lower the static air pressure exerted on the walls of the opening. The venturi thus creates a low-pressure area within the carburetor, figure 4-14.

The principle of the airfoil is also used to gain lower pressure conditions in the venturi. A fuel supply tube or jet is placed in the venturi section. The action of the incoming air causes a high pressure on the front of the jet but a very low pressure on the back of the jet. Gasoline is available in this jet and streams out of the jet because a low-pressure area has been created by the action of the venturi and the airfoil, and because the gasoline is under atmospheric pressure which is greater. Greater pressure pushes the gasoline out of the discharge jet into the airstream. See figure 4-15 for a simplified illustration of carburetor operation.

As gasoline streams into the airflow, it is mixed thoroughly with the air. The best mixture of gasoline and air is fourteen or fifteen parts of air to one part of gasoline, by weight. This air-to-gasoline ratio can be changed for different operating conditions; heavy load and fast acceleration require more gasoline (richer mixture). The needle valve is used to change this ratio; it controls the amount of gasoline that is available to be drawn from the discharge jet.

The throttle or "butterfly" is mounted on a shaft beyond the venturi section. The operator of the engine controls its setting to regulate the engine speed. When the throttle is wide open it does not restrict the flow of air; air flows easily through the carburetor and the engine is operating at its top speed. As the operator closes the throttle, the flow of air is restricted. A smaller amount of air can rush through the carburetor, therefore, the air pressures at the venturi section are not as low and less gasoline streams from the discharge holes. With less fuel mixture in the combustion chamber, the piston is pushed down with less force during combustion; power and speed are reduced.

The ratio of air to fuel remains about the same through the different throttle settings. However, when the throttle is closed and the

engine begins to idle, very little air is drawn through the carburetor and the difference between atmospheric pressure and venturi air is slight. Little gasoline is drawn from the discharge jet. In fact, the mixture of gasoline is so lean that a special idling device must be built into the carburetor to provide a richer mixture for idling.

In some carburetors, the main discharge jet is continued past the venturi section to the area of the throttle. It discharges fuel into a small well and jet that are behind the throttle when it is closed. The air pressure behind the throttle is very low. Therefore, gasoline streams from the idle jet readily and mixes with the small amount of air that is coming through the carburetor and a rich fuel mixture is provided for idling. A threaded needle valve called the *idle valve* controls the amount of gasoline that can be drawn from the idle jet. Many carburetors use the air bleed principle to help atomize the gasoline and to make the needle valve easier to adjust, figure 4-16.

When a cold engine is to be started, an extremely rich mixture of gasoline and air must be provided if the engine is to start easily. The choke provides this rich mixture, figure 4-17. The choke is a round disc placed in the air horn before the venturi section. For starting, the choke is closed, shutting off most of the carburetor's air supply. When the engine is turned over slowly, usually by hand, little air is drawn through the carburetor. The air pressure within the carburetor is very low and gasoline streams from the main discharge jet, mixing with the air that does get by the choke. As soon as the engine starts, the choke is opened. The fuel mixture provided when the engine is choked is so rich that all the gasoline may not vaporize with the air and raw liquid gasoline may be drawn into the combustion chamber. Continued operation

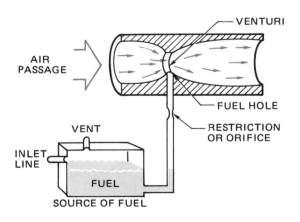

Fig. 4-15 Simplified carburetor operation

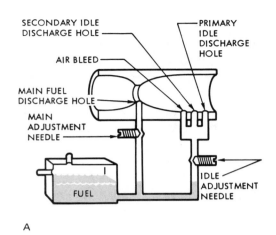

A

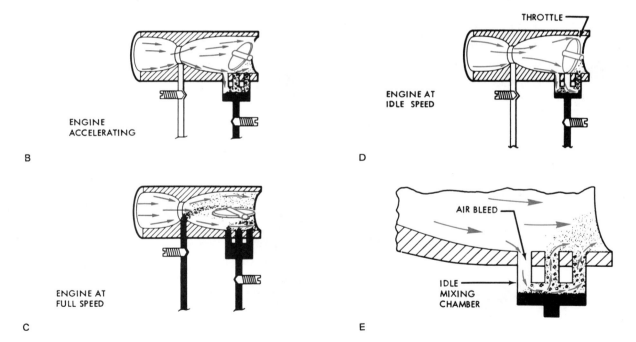

Fig. 4-16 (A) Schematic of basic carburetor parts, (B) engine accelerating, (C) engine at full speed, (D) engine at idle speed, and (E) air bleed principle

with the choke closed can cause crankcase dilution (the raw gasoline seeps into the crankcase diluting the lubricating oil).

The complete carburetor operation includes:

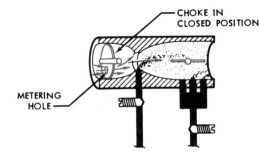

Fig. 4-17 The choke provides a rich mixture for starting.

- the float and float bowl maintaining a constant reservoir of gasoline in the carburetor
- the venturi section producing a low - pressure area
- the needle valve controlling the richness or leanness of fuel mixture
- the main discharge jet spraying gasoline into the airstream
- the idle valve providing a rich mixture for idling conditions
- the choke producing an extremely rich mixture for easy starting.

See figures 4-18 and 4-19.

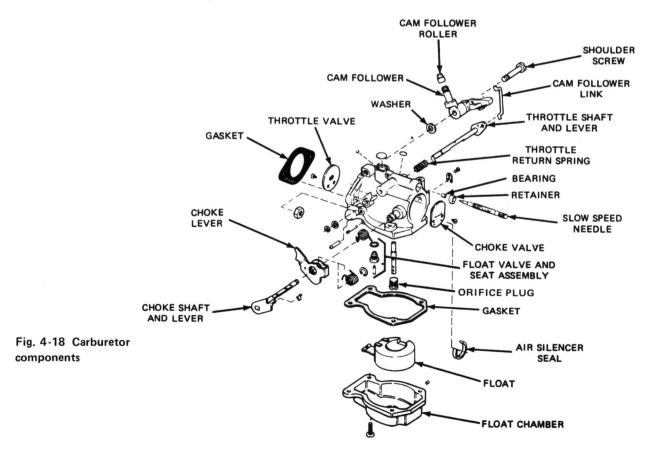

Fig. 4-18 Carburetor components

FUEL SYSTEM COMPONENTS
AND THEIR FUNCTION

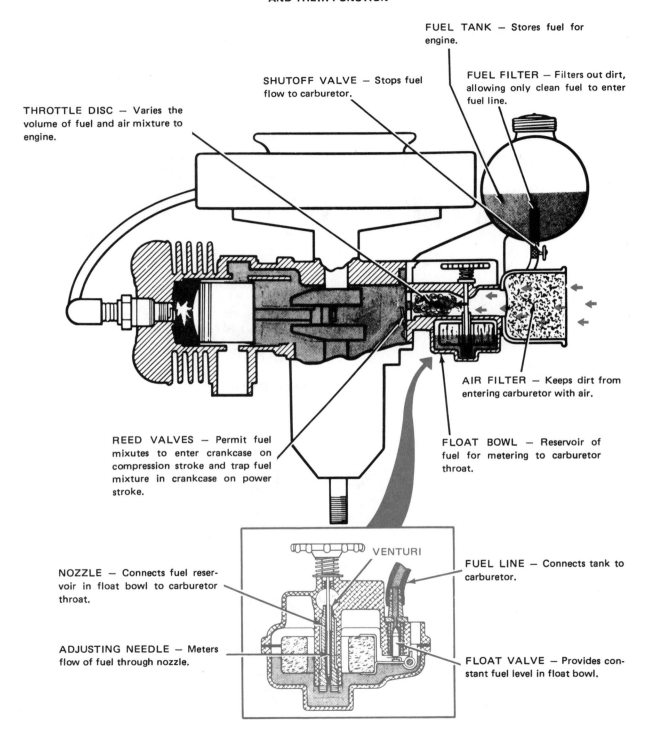

FUEL TANK — Stores fuel for engine.

FUEL FILTER — Filters out dirt, allowing only clean fuel to enter fuel line.

SHUTOFF VALVE — Stops fuel flow to carburetor.

THROTTLE DISC — Varies the volume of fuel and air mixture to engine.

AIR FILTER — Keeps dirt from entering carburetor with air.

REED VALVES — Permit fuel mixutes to enter crankcase on compression stroke and trap fuel mixture in crankcase on power stroke.

FLOAT BOWL — Reservoir of fuel for metering to carburetor throat.

VENTURI

NOZZLE — Connects fuel reservoir in float bowl to carburetor throat.

FUEL LINE — Connects tank to carburetor.

ADJUSTING NEEDLE — Meters flow of fuel through nozzle.

FLOAT VALVE — Provides constant fuel level in float bowl.

Fig. 4-19 Two-stoke cycle carburetor and fuel system

Fig. 4-20 Chain saws commonly use diaphragm carburetors

Primer

It should be noted that some carburetors have a *primer* instead of a choke for providing the extra gasoline and rich mixture for starting. Pushing the primer causes liquid or raw gasoline to be squirted into the venturi section of the carburetor. One or two pushes on the primer provides enough gasoline for starting the engine.

DIAPHRAGM CARBURETOR

The diaphragm carburetor has come into wide use, especially on chain saws. It is also found on other applications where the engine may be tipped at extreme angles. The diaphragm supplies the carburetor with a constant supply of gasoline.

The diaphragm carburetor may be gravity fed. Crankcase pressure moves the diaphragm and associated linkages to allow gasoline to enter the fuel chamber. The gasoline is available by gravity and the in-and-out motion of the diaphragm meters out fuel in the correct amount.

Diaphragm carburetors are also made with a built-in fuel pump, figures 4-21 and 4-22. The fuel pump operates on differences in crankcase pressure; its action is much the same as that of the outboard fuel pump previously discussed. Diaphragm carburetors are widely used on small engines.

Other Carburetors

A common carburetor variation uses both the fuel pump and suction feed principles. The Briggs and Stratton Pulsa-Jet® is such a carburetor. At first this carburetor may appear to be an ordinary suction type carburetor but closer examination shows there are two fuel pipes into the carburetor — one long and one short.

The low pressure created by the intake stroke activates the fuel pump drawing gasoline through the long pipe and discharging it into a smaller constant level fuel chamber at the top of the tank. This provides a constant level of gasoline regardless of the amount of gasoline in the tank. It also reduces the amount of lift

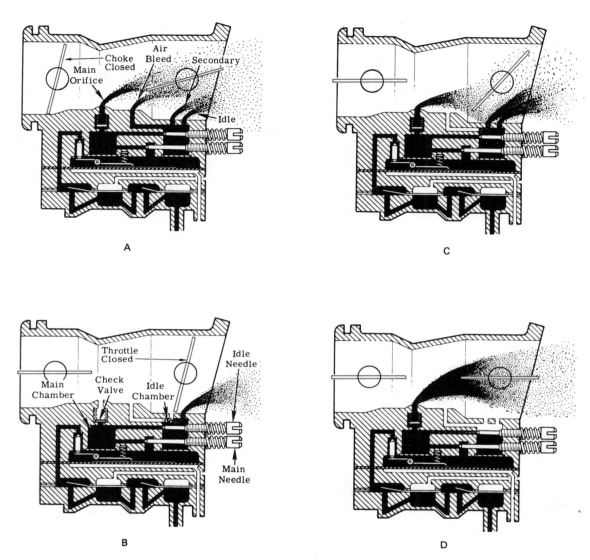

Fig. 4-21 Diaphragm carburetor with built-in fuel pump. (A) choke closed, (B) idle, (C) part throttle, (D) full throttle

required to deliver gasoline into the venturi. This carburetor permits a larger venturi and improves the engine's horsepower rating. The Briggs and Stratton Pulsa-Jet® is shown in figures 4-23 and 4-24.

AIR BLEEDING CARBURETORS

There is a tendency for carburetors to supply too rich a fuel mixture at high speeds; the ratio of gasoline to air increases as the velocity

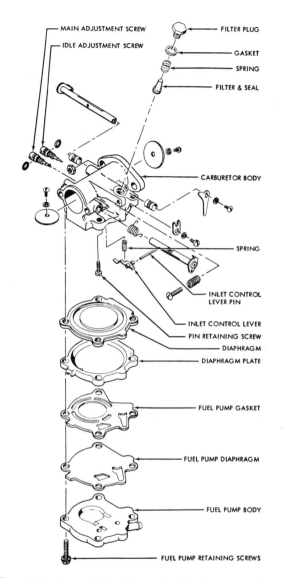

Fig. 4-22 An exploded view of a diaphragm carburetor with a built-in fuel pump

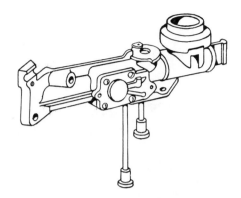

Fig. 4-23 Pulsa-Jet® carburetor

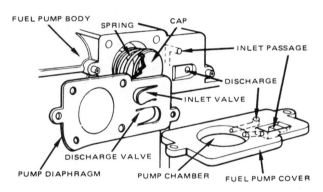

Fig. 4-24 Pulsa-Jet® carburetor pump

of the air passing through the carburetor increases. One method that is commonly used to correct this condition is *air bleeding* (see figure 4-16, view E). Air bleeding involves introducing a small amount of air into the main discharge well vent to restrict the flow of gasoline from the main discharge jet. As engine speeds are increased, greater amounts of air are brought into the main discharge well vent placing a greater restriction on the gasoline flow. The additional air overcomes the carburetor's natural tendency to provide too rich a mixture at high speeds. This action maintains the proper ratio of fuel and air between a throttle setting of one-fourth to wide open. The air that enters the discharge well vent mixes with the gasoline and is drawn through the main discharge jet into the main airstream. The Zenith® carbu-

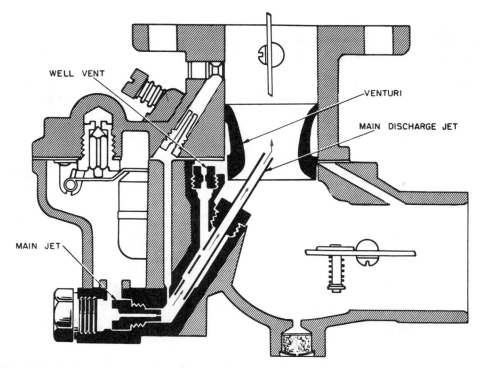

Fig. 4-25 Air brought into the well vent bleeds into the main discharge jet, maintaining the correct air-fuel ratio throughout throttle range.

retor shown in figure 4-25, uses the principle of air bleeding.

ACCELERATING PUMP

Another problem inherent in all carburetors is a response lag when the throttle is quickly opened. Air can react very quickly to an increased demand but gasoline lags behind. The result is too lean a mixture and slow acceleration. Carburetors equipped with an accelerating pump provide instant response for rapid acceleration.

The main parts of the accelerating pump are the spring, vacuum piston, and fuel cylinder, figure 4-26. At idling and low operating speeds, the vacuum piston is drawn to the top

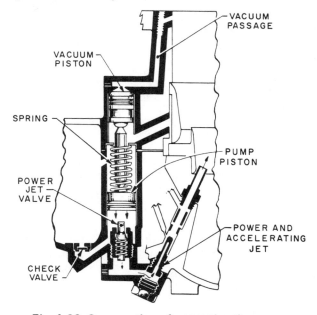

Fig. 4-26 Cross section of an accelerating pump

of the fuel cylinder by the engine vacuum. This vacuum is strong enough to overcome the force of the spring holding the piston at the top of its stroke. Now the fuel cylinder is filled with gasoline.

If the throttle is suddenly opened, the engine vacuum drops enough for the piston spring to overcome the force of the vacuum pushing the pump piston down the fuel cylinder. This reserve amount of gasoline in the fuel cylinder is forced into the main discharge jet and on into the carburetor venturi.

CARBURETOR ADJUSTMENTS

The engine manufacturer sets the carburetor adjustments at the factory. These settings cover normal operation. However, after a long period of usage, or under special operating conditions, it may be necessary to adjust the carburetor. In adjusting the carburetor, the main needle valve and the idle valve are both reset to give the desired richness or leanness of fuel mixture. Too lean a mixture can be detected by the engine missing and backfiring. Too rich a mixture can be detected by heavy exhaust and sluggish operation.

Procedure

This is a typical procedure for adjusting a carburetor for maximum power and efficiency.

1. Close the main needle valve and idle valve finger tight. Excessive force can damage the needle valve. Turn clockwise to close.
2. Open the main needle valve one turn. Open the idle valve 3/4 turn. Turn counterclockwise to open.
3. Start the engine, open the choke, and allow the engine to reach operating temperature.

4. Run the engine at operating speed (2/3 to 3/4 of full throttle). Turn the main needle valve clockwise until the engine slows down indicating too lean a mixture. Note the position of the valve. Turn the needle valve counterclockwise until the engine speeds up and then slows down indicating too rich a mixture. Note the position of the valve. Reposition the valve halfway between the rich and lean settings.
5. Close the throttle so the engine runs slightly faster than normal idle speed. Turn the idle valve clockwise until the engine slows down, then turn the idle valve counterclockwise until the engine speeds up and idles smoothly. Adjust the idle-speed regulating screw to the desired idle speed, figure 4-27.

NOTE: Idle speed is not the slowest speed at which the engine can operate; rather, it is a slow speed that maintains good airflow for cooling and a good take-off spot for even acceleration. A

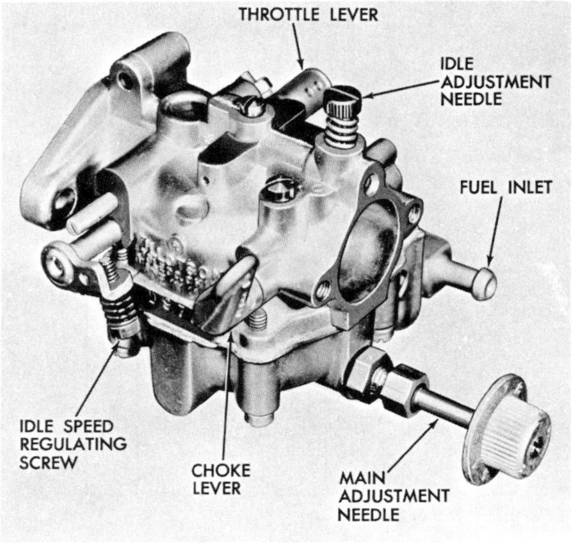

Fig. 4-27 Float-feed type carburetor

tachometer and the manufacturer's specifications regarding proper idle speed are necessary for the best adjustment.

6. Test the acceleration of the engine by opening the throttle rapidly. If acceleration is sluggish, a slightly richer fuel mixture is usually needed.

CARBURETOR ICING

Carburetor icing can occur when the engine is cold and certain atmospheric conditions are present, figure 4-28. If the temperature is between 28°F and 58°F and the relative humidity is above 70 percent, carburetor icing can take place.

Fuel mixture of gasoline and air rushes through the carburetor and the rapid action of evaporating gasoline chills the throttle plate to about 0°F. Moisture in the air condenses and freezes on the throttle plate when the relative humidity is high, figure 4-29. This formation of ice restricts the airflow through the carburetor; at low or idle settings the ice can completely block off the airflow stalling the engine.

When this condition is present, the engine can be restarted but stalls again at low or idle speeds. As soon as the carburetor is warm enough to prevent ice formation, normal operation can take place.

VAPOR LOCK

Vapor lock can occur anywhere along the fuel line, fuel pump, or in the carburetor when temperatures are high enough to vaporize the gasoline. Gasoline vapor in these places cuts off the liquid fuel supply stalling the engine. If vapor lock occurs, the operator must wait until the carburetor, gas line, and fuel pump cool off and the gasoline vapor returns to liquid before the engine will restart. Vapor lock usually occurs on unseasonably hot days and is more troublesome at high altitudes.

GASOLINE

Most internal combustion engines burn gasoline as their fuel. Gasoline comes from petroleum, also called crude oil. Crude oil is actually a mixture of different hydrocarbons such as gasoline, kerosene, heating oil, lubricating oil, and asphalt. These chemicals are

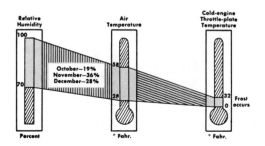

Fig. 4-28 Humidity and temperature conditions which may lead to carburetor icing.

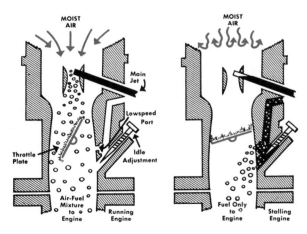

Fig. 4-29 Under carburetor icing conditions, ice forms on the throttle plate cutting off air to the engine when the throttle closes.

all hydrocarbons but they have characteristics that are quite different. Hydrogen is a light, colorless, odorless gas; carbon is black and solid. Different combinations of carbon and hydrogen give the different characteristics of hydrocarbon products.

The various hydrocarbons are separated by distillation of the crude oil. Crude oil is first heated to a temperature of 700° to 800°F and then released into a fractionating or bubble tower, figure 4-30. The tower contains 20 to 30 trays through which hydrocarbon vapors can rise from below. When the heated crude oil is released into the tower, most of it flashes into vapor. The vapors cool as they rise. Each type of hydrocarbon condenses at a different tray level. Heavier hydrocarbons condense first at relatively high temperatures. The lighter hydrocarbons, such as gasoline, condense high in the tower at relatively low temperatures. Gasoline is then further processed to improve its qualities.

Good gasoline must have several characteristics.

- It must vaporize at low temperatures for good starting.
- It must be low in gum and sulfur content.
- It must not deteriorate during storage.
- It must not knock in the engine.
- It must have proper vaporizing characteristics for the climate and altitude.
- It must burn cleanly to reduce air pollution.

The vaporizing ability of gasoline is the key to its success as a fuel. In order to burn inside the engine, the hydrocarbon molecules of the gasoline must be mixed with air since oxygen is also necessary for combustion. To mix the gasoline and air, there must be rapid motion

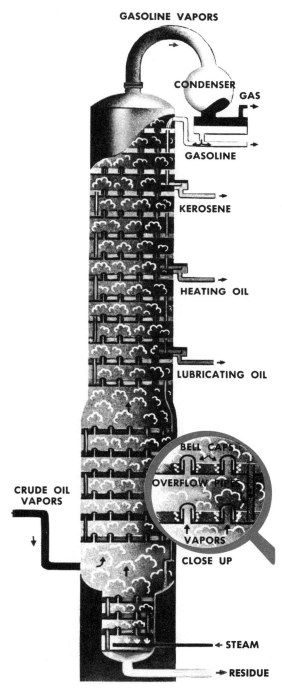

Fig. 4-30 Distillation of petroleum in the bubble tower

Petroleum fuel — a mixture of many hydrocarbon compounds of different weights.

Like a stack of gravel, consisting of many pebbles basically alike, but of many different sizes—from larger, heavier ones to particles of fine dust.

Extreme turbulence is required to maintain suspension of all particles of fuel vapor — rapid movement through the carburetor and manifold, turbulence in the crankcase and combustion chamber.

Like a whirlwind of gravel on a country road, all particles remain suspended as long as agitation and motion continue.

Heavier particles of the fuel settle or rain out of the fuel vapor with lessening turbulence, to a puddle below.

The very lightest particles (hydrocarbons) remain suspended like the heavier pebbles falling out of the whirlpool and returning to the roadway with lessening of the whirling wind. The very lightest particles remain suspended as dust.

Fig. 4-31 Turbulence is necessary to keep the gasoline molecules suspended in the air.

and turbulence to keep the molecules suspended in the air, figure 4-31. There is rapid motion and turbulence in the carburetor; this turbulence continues into the combustion chamber. Combustion chambers are designed to create the maximum turbulence so that gasoline molecules will stay suspended in the air until they are ignited. A thorough mixture ensures smooth and complete burning of gasoline which delivers maximum power.

Although the vaporizing ability of gasoline makes it an excellent fuel, it also presents a great fire and explosion hazard. When vaporized, a gallon of gasoline produces 21 cubic feet of vapor. If this vapor were combined with air in a mixture of 1.4 to 7.6 percent gasoline by volume, it would explode when ignited. Therefore, only one gallon of gasoline properly vaporized would completely fill an average living room with explosive vapor. The safety rules given in unit 2 for the use and storage of gasoline should always be followed.

In engines, the compression of the fuel mixture of gasoline and air results in high combustion pressures and the force necessary to move the piston. This compression may also cause the gasoline to explode (*detonate*) in the engine instead of burning smoothly.

Detonation is also called knocking, fuel knock, spark knock, carbon knock, and ping. To understand it, one has to think in slow motion. Refer to figure 4-32. The spark plug ignites the fuel mixture, and a flame front moves out from this starting point. As the flame front sweeps across the combustion chamber, heat and pressure build. The unburned portion of the fuel mixture ahead of the flame front is exposed to this heat and pressure. If it self-detonates, two flame fronts are created which race toward each other. The last unburned portion of fuel caught between the two fronts explodes with

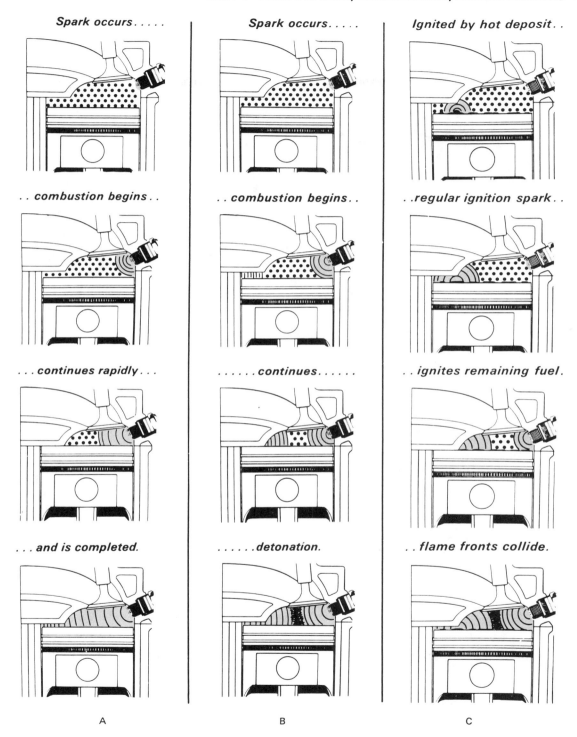

Fig. 4-32 (A) Normal combustion, (B) detonation and (C) preignition

Fig. 4-33 Detonation damage

Fig. 4-34 Preignition damage

hammerlike force. Detonation causes a knocking sound in the engine and power loss. Repeated detonation can damage the piston, figure 4-33.

Preignition causes the same undesirable effects as detonation, including engine damage, figure 4-34. The cause of preignition, however, is somewhat different. Hot spots, or red-hot carbon deposits, actually ignite the fuel mixture and begin combustion before the spark plug fires.

The antiknock quality of a gasoline, or its ability to burn without knocking, is called its *octane rating.* Gasoline that has no knocking characteristics at all is rated at 100. Usually, tetraethyl lead is added to gasoline to give it the proper antiknock quality. If lead is not used as in unleaded gasoline, special aromatic compounds are used to obtain smooth burning.

The higher the compression ratio of the engine, the higher the octane requirements for the engine's fuel. Low-grade gasoline with an octane rating of 70-85 is suitable for compression ratios of 5-7 to 1. Regular grade gasoline with an octane rating of 88-94 is

suitable for compression ratios of 7-8.5 to 1. Most small engine manufacturers recommend the use of regular gasoline. High-octane premium gasoline, leaded or unleaded, does not improve the performance of small engines. Premium gasoline with an octane rating of about 100 is suitable for compression ratios of 9-10 to 1. Super premium has an octane rating of over 100 and is good for engines with compression ratios of 9.5-10.5 to 1.

ENGINE SPEED GOVERNORS

Speed governors are used to keep engine speed at a constant rate regardless of the load. For instance, a lawn mower is required to cut tall as well as short grass. A governor ensures that the engine operates at the same speed in spite of the varying load conditions. Engines on many other applications also need this constant speed feature.

Speed governors are also used to keep the engine operating speed below a given, preset rate, so that the engine speed will not surpass this rate. This maximum rate is established and the governor set accordingly by the engineers at the factory. This type of speed governor protects both the engine and the operator from speeds that are dangerously high.

In the case of lawn mowers the governor also limits the top speed to conform to federal safety standards which limit the blade tip speed to 19,000 feet per minute.

There are two main governor systems used on small gasoline engines: (1) mechanical or centrifugal type and (2) pneumatic or air vane type. Although there are other types of governors, most of the governors used on small engines are included in one of these categories.

MECHANICAL GOVERNORS

The mechanical governor operates on centrifugal force, figures 4-35 to 4-37. With this method, counterweights mounted on a geared shaft, a governor spring, and the associated governor linkages keep the engine speed at the desired rpm. For constant speed operation the action follows this pattern: when the engine is stopped, the mechanical governor's spring pulls the throttle to an open position. The governor spring tends to keep the throttle open but as the engine speed increases centrifugal force throws the hinged counterweights further and further from their shaft. This action puts tension on the spring in the other direction to close the throttle.

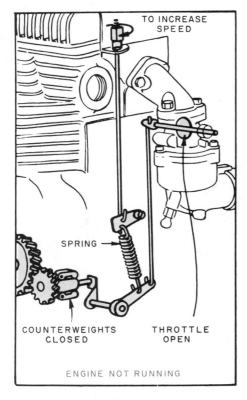

Fig. 4-35 A mechanical governor which operates on centrifugal force

Fig. 4-36 Mechanical governor (centrifugal)

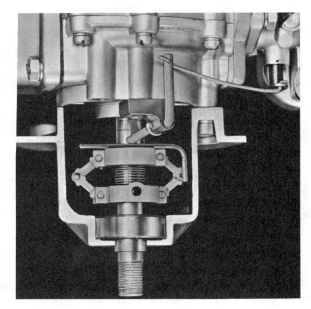

Fig. 4-37 Mechanical governor

At governed speed, the spring tension is overcome by the counterweights and the throttle opens no further. A balanced position is maintained and the engine assumes its maximum speed.

If, however, a greater load is placed on the engine, it slows down; the hinged counterweights swing inward due to lessened centrifugal force; the governor spring becomes dominant, opening the throttle wider. With a wider throttle opening, the engine speeds up until governed speed is reached again. The action described is fast and smooth; little time is needed for the governor to meet revised load conditions.

The governed speed can be changed somewhat by varying the tension on the governor spring. The more tension there is on the governor spring, the higher the governed speed.

Another commonly used centrifugal or mechanical governor is the flyball type. With this governor, round steel balls in a spring loaded raceway move outward as the engine speed increases. The centrifugal force applied by these balls increases until it balances the spring tension holding the throttle open. At this point, the governed speed is reached.

PNEUMATIC OR AIR VANE GOVERNORS

Many engines have a pneumatic or air vane governor, figures 4-38 through 4-40. Again, the governor holds the throttle at the governed speed position which prevents it from opening further. An air vane, which is located near the flywheel blower, controls the speed. As engine speed increases, the flywheel blower pushes more air against the air vane causing it to change position. At governed speed the air vane position overcomes the governor spring tension and a balance is assumed.

Engines are often designed so that they can be accelerated freely from idle speed up to governed speed. With these engines, the governor spring takes on very little tension at low speeds. The operator has full control. As the throttle is opened further and further, there is more tension on the spring, both from the operator and from the strengthening governor action. At governed speed, the force to close the throttle balances the spring tension to open the throttle.

It should be noted that not all engines have engine speed governors. Engines that are always operated under a near constant load do not need the governor. The load of an outboard engine is a good example of a near constant load. Although different loads are placed on the outboard (number of people in boat, position of people in boat, use of trolling mechanism to decrease boat's speed, etc.), the load is constant at full throttle for a given period of use.

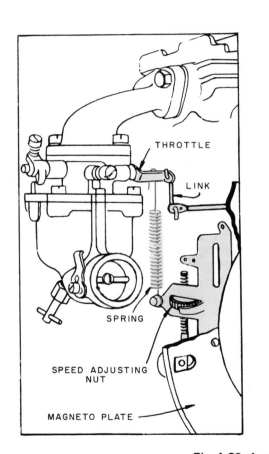
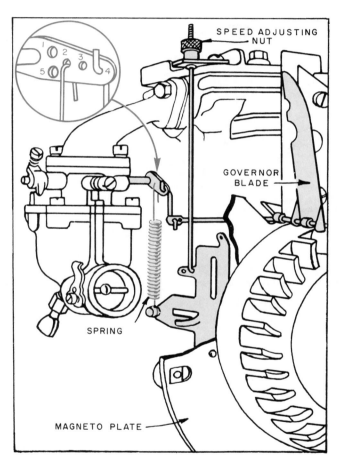

Fig. 4-38 A pneumatic or air vane governor

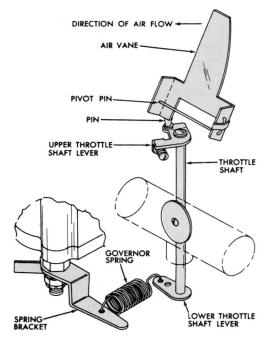

DIRECTION OF AIR FLOW

AIR VANE

PIVOT PIN

PIN

UPPER THROTTLE SHAFT LEVER

THROTTLE SHAFT

GOVERNOR SPRING

SPRING BRACKET

LOWER THROTTLE SHAFT LEVER

Fig. 4-39 A pneumatic or air vane governor

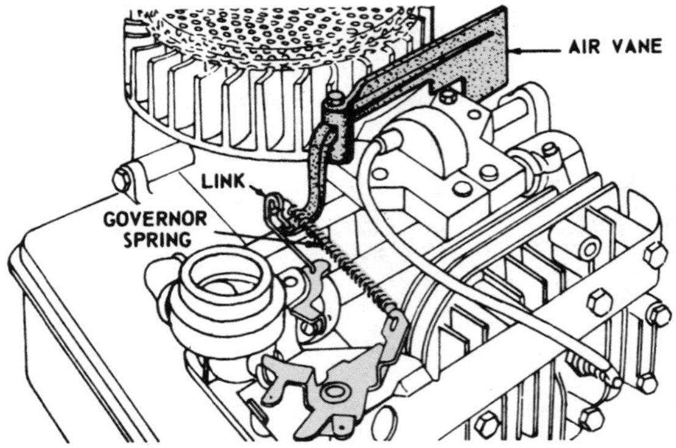

AIR VANE

LINK

GOVERNOR SPRING

Fig. 4-40 Air vane governor

REVIEW QUESTIONS

1. Name the basic parts of a typical fuel system.

2. What force is used to operate most outboard fuel pumps?

3. How can a pressurized fuel system be recognized?

4. Briefly describe the task of the carburetor.

5. Name the main parts that make up the carburetor.

6. What causes air to flow through the carburetor?

7. What causes low air pressure in the venturi section of the carburetor?

8. What is the function of the throttle?

9. What is the function of the choke?

10. What is the function of the needle valve?

11. Explain the action of the float in a float-type carburetor.

12. What is the advantage of a diaphragm carburetor?

13. Explain the principle of air bleeding.

14. Explain the operation of the accelerating pump.

15. What is a rich mixture?

16. Why is turbulence important inside the combustion chamber?

17. What two main purposes do engine speed governors serve?

18. What are the two main governor systems used on small gasoline engines?

19. Are engines made that do not have engine speed governors?

20. Can the governed speed be changed? How?

CLASS DEMONSTRATION TOPICS

- Using a fully assembled engine, have the students trace the flow of fuel and also identify the parts of the fuel system.

- Demonstrate with an engine operating how to set a carburetor for maximum power.

- Demonstrate with an operating engine how a too lean or too rich mixture affects operation and acceleration.

- Demonstrate the action of a carburetor with an atomizer or insect sprayer.

- Disassemble and inspect a fuel pump (automotive or outboard).

- With an operating engine, demonstrate how the governor maintains governed speed and does not allow the throttle to open to "full."

- Demonstrate how governed speed can be changed. Stop the engine to readjust the governor.

LABORATORY EXPERIENCE 4-1
BASIC PARTS OF THE CARBURETOR

The carburetor is a key part of the gasoline engine. By mixing gasoline and air together, it provides the engine with the explosive fuel mixture necessary for ignition and power. Its main parts allow the engine to be operated under different load conditions, different speed conditions, and different starting conditions.

STUDENT ASSIGNMENT

You are to remove the carburetor from the engine, then: (1) disassemble the carburetor, (2) study the basic parts, and (3) reassemble the carburetor. When you have reassembled the carburetor, reinstall it on the engine. It is essential that you record your work in a work record box as you complete each step of the disassembly procedure.

Disassembly Procedure

Instructors may want to supplement or revise specific steps of this procedure since there are many makes of carburetors. The following disassembly procedure is a general guide.

1. Remove the metal shrouding to expose the carburetor, if necessary.

2. Close the fuel shutoff valve at the gas tank or drain the gas tank.

3. Drain the carburetor float bowl, if the drain valve is on the float bowl.

4. Remove the air cleaner. (Use care if it contains oil.)

5. Remove the gas line from the carburetor.

6. Remove the throttle and governor connections.

7. Remove the carburetor from the engine.

8. Disassemble the float bowl and inspect the float chamber.

9. Remove the main or high-speed needle valve (counterclockwise).

10. Remove the idle screw or slow-speed needle valve (counterclockwise).

CAUTION: Never force needle valves in against their seats — they will be damaged. Do not remove the choke or throttle valves.

WORK RECORD BOX

PART	DISASSEMBLY (nuts, bolts, etc.)	OPERATION PERFORMED	TOOL USED

REVIEW QUESTIONS

Study the carburetor and answer the following questions.

1. Does the engine have a fuel pump?

2. Is the carburetor suction fed?

3. Is the carburetor gravity fed?

4. Does the carburetor have a float bowl?

5. Does the carburetor have a needle valve?

6. Does the carburetor have an idle valve?

7. Does the carburetor have a choke?

8. Does the carburetor have an air cleaner?

9. What does the carburetor do?

10. What causes air to flow through the carburetor?

11. What is the function of the venturi section?

12. What is the function of the throttle?

13. What is the function of the choke?

14. Trace the path of fuel and fuel mixture through the engine.

LABORATORY EXPERIENCE 4-2
ADJUSTING THE CARBURETOR FOR MAXIMUM POWER AND EFFICIENCY

The correct carburetor adjustment is vital to top engine performance. An improper mixture of fuel and air can cause sluggish operation, overheating, increased part wear, excessive fuel consumption, fouled spark plugs, and other troubles. The needle valve and idle valve control the ratio of gasoline to air, therefore, the valves must be correctly positioned. Normal operation uses a ratio of about 15 parts air to 1 part gasoline (by weight) although the ratio may be richer or leaner to suit conditions.

STUDENT ASSIGNMENT

The carburetor is adjusted for maximum power and efficiency. These adjustments should be made with the engine operating and with the carburetor and engine at normal operating temperature.

> CAUTION: Do not operate the engine without proper ventilation or exhaust system. Also, be certain that no loose clothing can become caught in the moving parts.

Adjustment Procedure

Instructors may want to supplement or revise specific steps of this procedure since there are many makes of carburetors and engines. The following adjustment procedure is a general guide.

1. Close the main needle valve and idle valve finger tight. Excessive force can damage the needle valves. Turn clockwise to close.

2. Open the main needle valve one turn. Open the idle valve 3/4 turn. Turn counterclockwise to open.

3. Start the engine, open the choke and allow the engine to reach operating temperature.

4. Run the engine at operating speed (2/3 to 3/4 of full throttle). Turn the main needle valve clockwise until the engine slows down indicating too lean a mixture. Note the position of the valve. Turn the

needle valve counterclockwise until the engine speeds up and then slows down indicating too rich a mixture. Note the position of the valve. Reposition the valve halfway between the rich and lean setting.

5. Close the throttle so the engine runs slightly faster than normal idle speed. Turn the idle valve clockwise until the engine slows down. Then turn the idle valve back out until the engine speeds up and idles smoothly. Adjust the idle-speed regulating screw to the desired idle speed.

   NOTE: Idle speed is not the slowest speed at which the engine can operate. Rather, it is a slow speed that maintains good airflow for cooling and a good takeoff spot for even acceleration. A tachometer and the known proper idle speed are necessary for the best adjustment.

6. Test the acceleration of the engine by opening the throttle rapidly. If acceleration is sluggish, a slightly richer fuel mixture is probably necessary.

REVIEW QUESTIONS

Adjust the carburetor and answer the following questions.

1. What problem does too lean a fuel mixture cause? Too rich?
2. Describe the engine exhaust produced by too rich a fuel mixture.
3. Explain how needle and idle valves can be damaged.

LABORATORY EXPERIENCE 4-3
FUEL PUMPS

The diaphragm fuel pump action is simple and reliable. The flexible diaphragm moves out and gasoline enters the fuel chamber through the inlet check valve. Now the diaphragm moves in and the fuel is forced onto the carburetor through the discharge check valve. Each valve allows fuel flow in one direction only. In outboard fuel pumps the diaphragm is operated by the differences of crankcase pressure. In the automobile fuel pump the diaphragm is operated by cam action.

NOTE: Outboard fuel pumps are small, fairly intricate and somewhat delicate. Their disassembly and inspection is not recommended for beginners. The automobile fuel pump construction and action is basically the same and these pumps are larger and easier for the beginner to handle. Discarded automobile fuel pumps are also easy to obtain.

STUDENT ASSIGNMENT

You are to disassemble and inspect the fuel pump. Be sure to note how the diaphragm can move in and out. Also, note how the inlet and discharge valves allow fuel flow in only one direction. Trace the flow of fuel through the pump. It is essential that you record your work in a work record box as you complete each step of the disassembly procedure.

Disassembly Procedure

Instructors may want to supplement or revise specific steps of this procedure since students may be working on either an automotive or outboard fuel pump. The following disassembly procedure is a general guide.

1. Remove fuel lines from the tank and to the carburetor.
2. Remove the fuel pump from the engine.
3. Remove machine screws that hold the pump together at the diaphragm.
4. Carefully separate the halves of the pump, exposing the diaphragm.
5. Remove and inspect the inlet and discharge valves (on some pumps this is not possible).

WORK RECORD BOX

PART	DISASSEMBLY (nuts, bolts, etc.)	OPERATION PERFORMED	TOOL USED

Reassembly Procedure

Reverse the disassembly procedure.

REVIEW QUESTIONS

Study the fuel pump and answer the following questions.

1. Explain what causes the diaphragm to move in and out.

2. Describe the shape of the inlet and discharge valves.

LABORATORY EXPERIENCE 4-4
AIR VANE GOVERNOR (DISASSEMBLE AND INSPECT)

The air vane governor is operated by the flywheel air flow which is channeled by it. The faster the engine speed the more the hinged vane is deflected by the air blast. The governor spring works to hold the throttle wide open while the air vane works to close the throttle. Governed speed is reached when the force to open the throttle and the force to close the throttle balance each other. The greater the governor spring tension the higher the governed speed since a bigger air blast from the flywheel is needed to balance the spring tension.

STUDENT ASSIGNMENT

You are to examine the air vane type of engine speed governor in order to identify and become familiar with its various parts. It is essential that you record your work in a work record box as you complete each step of the disassembly procedure.

Disassembly Procedure

Instructors may want to supplement or revise specific steps of this procedure since there are many makes of engines. The following disassembly procedure is a general guide.

1. Remove all air shrouding that covers the flywheel area. Expose the engine speed governor.

2. Study the air scoops on the flywheel, the governor spring, the air vane, and the associated linkages.

3. Determine how the tension on the governor spring can be changed to increase or decrease governed speed.

4. Remove the governor parts only if directed to do so by the instructor.

WORK RECORD BOX

PART	DISASSEMBLY (nuts, bolts, etc.)	OPERATION PERFORMED	TOOL USED

Reassembly Procedure

Reverse the disassembly procedure. Use great care with the delicate parts. The parts must not bind at any point; free movement is essential for proper operation.

REVIEW QUESTIONS

Study the air vane governor and answer the following questions.

1. What are the main parts of the air vane governor?

2. If the air shrouding were removed from the engine would the governor operation be affected? Why?

3. How can the governor spring tension be changed on the engine?

LABORATORY EXPERIENCE 4-5
ENGINE SPEED GOVERNOR ACTION

The engine speed governor on most small engines is provided to enable the engine to deliver constant rpm regardless of engine load. The engine operating under light load will have quite a bit of reserve throttle opening to enable the engine to maintain constant speed when load is increased. The governed speed can usually be adjusted through a range of several hundred rpm to satisfy the particular needs of the engine operator. To increase the governed speed the tension on the governor spring is increased; to decrease speed the spring tension is decreased. On some engines the governor spring is hooked into numbered holes. On other engines a nut on a threaded link to the spring is tightened or a lever arm is repositioned. Many systems are used.

CAUTION: Do not operate the engine without a proper ventilation or exhaust system. Also, be certain that no loose clothing can become caught in the moving parts.

STUDENT ASSIGNMENT

You are to study the engine speed governor, either mechanical or air vane type, to become familiar with the action of the governor on the engine throttle. As a part of this study you will learn how to adjust the governor for the various speed settings. It is essential that you record your work in a work record box as you complete each step of the procedure.

Procedure

Either mechanical or air vane governors can be used for this laboratory experience. Use an engine that is on a test stand, not one connected to an implement.

1. With the engine stopped, work the throttle and observe the throttle positions for idle and wide open.

2. Start the engine bringing the engine to governed speed (many governors do this automatically). Now note the throttle opening. Using a tachometer on the end of the crankshaft, find the engine rpm. Record rpm in a work record box.

3. Stop the engine.

4. Readjust the governor spring tension for increased top speed or readjust the governor spring tension for decreased top speed.

5. Start the engine bringing the engine to new governed speed. Note the throttle opening. Using a tachometer on the end of the crankshaft, find the engine rpm. Record rpm in a work record box.

6. Stop the engine.

WORK RECORD BOX

GOVERNOR SETTING	RPM

REVIEW QUESTIONS

Upon completion of the adjustment procedure, answer the following questions.

1. Explain the function of the engine speed governor.

2. How was the governor spring tension adjusted on the engine?

3. Was the engine speed governor a mechanical or air vane governor?

4. How does the throttle opening in step 5 of the procedure compare with that in step 2?

UNIT 5

Lubrication

OBJECTIVES   After completing this unit, the student should be able to:
- Discuss the several functions of engine lubricating oil.
- Discuss friction bearings and antifriction bearings.
- Discuss lubricating oil quality designations and SAE numbers.
- Discuss how a two-stroke cycle engine is lubricated.
- Correctly drain and refill the crankcase with oil.
- Explain how a two-stroke cycle engine is lubricated.

Whenever surfaces move against one another they cause friction which results in heat and wear. Lubricating oils have one main job to perform in the engine — to reduce friction. The lubricating oil provides a film that separates the moving metal surfaces and keeps the contact of metal against metal to an absolute minimum, figure 5-1.

Without a lubricating oil or with insufficient lubrication, the heat of friction builds up rapidly. Engine parts become so hot that they fail. That is when the metal begins to melt, bearing surfaces *seize* (adhere to other parts), parts warp out of shape, or parts actually break. The common expression is to say the engine burns up.

As an example of friction, lay a book on the table and then push the book slowly. Notice the resistance. Friction makes the book difficult to slide. Now place three round

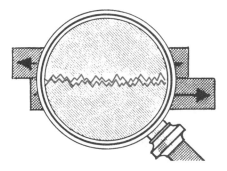

Fig. 5-1 Exaggerated view of metal surfaces in contact

pencils between the book and the table top and push the book. Notice how easily it moves. Friction has been greatly reduced. Oil molecules correspond to the pencils by forming a coating between two moving surfaces. With oil, the metal surfaces roll along on the oil molecules and friction is greatly decreased.

Besides reducing friction and the wear and heat it causes, the lubricating oil serves several other important functions.

Oil Seals Power.   The oil film seals power, particularly between the piston, piston rings, and cylinder walls. The great pressures in the combustion chamber cannot pass by the airtight seal which the oil film provides. If this oil film fails, a condition called blow-by exists. Combustion gases push by the film and enter the crankcase. Blow-by not only reduces engine power but also has a harmful effect on the oil's lubricating quality.

Oil Helps to Dissipate Heat.   Oil helps to dissipate heat by providing a good path for heat transfer. Heat conducts readily from inside metal parts through an oil film to outside metal parts that are cooled by air or water. Also, heat is carried away as new oil arrives from the crankcase and the hot oil is washed back to the crankcase.

Oil Keeps the Engine Clean.   Oil keeps the engine clean by washing away microscopic pieces of metal that have been worn off moving parts. These tiny pieces settle out in the crankcase or they are trapped in the oil filter, if one is used.

Oil Cushions Bearing Loads.   The oil film has a cushioning effect since it is squeezed from between the bearing surfaces relatively slowly. It has a shock absorber action. For example,

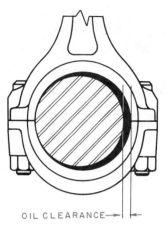

OIL CLEARANCE→

Fig. 5-2 Oil clearance between journal and bearing (exaggerated)

when the power stroke starts, the hard shock of combustion is transferred to the bearing surfaces; the oil film helps to cushion this shock. Figure 5-2 illustrates the oil clearance between the journal and the bearing.

Oil Protects Against Rusting.   The oil film protects steel parts from rusting. Air, moisture, and corrosive substances cannot reach the metal to oxidize or corrode the surface.

FRICTION BEARINGS

There are three types of friction bearings used in a small gasoline engine: journal, guide, and thrust, figure 5-3. The *journal bearing* is the most familiar. This bearing supports a revolving or oscillating shaft. The connecting rod around the crankshaft and the main bearings are examples. The *guide bearing* reduces the friction of surfaces sliding longitudinally against each other, such as the piston in the cylinder. The *thrust bearing* supports or limits the longitudinal motion of a rotating shaft.

Bearing inserts are commonly used in larger engines or heavy-duty small engines, particu-

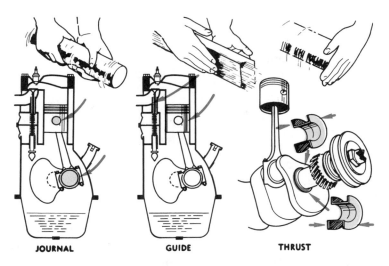

Fig. 5-3 Three major types of friction bearings

JOURNAL   GUIDE   THRUST

larly on the connecting rod and cap and the main bearings. The inserts are precision-made of layers of various metals and alloys. Alloys such as babbit, copper-lead, bronze, aluminum, cadmium, and silver are commonly used. Crankshaft surfaces are made of steel. Most small engines, however, do not have bearing inserts. They use only the aluminum connecting rod around the steel journal of the crankshaft.

ANTIFRICTION BEARINGS

Antifriction bearings are also commonly used in engines, figure 5-4. They substitute rolling friction for sliding friction. Ball bearings, roller bearings and needle bearings are of this type, figure 5-5. On many small engines, the main bearings are of the antifriction type.

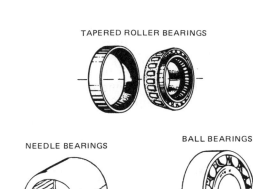

TAPERED ROLLER BEARINGS

NEEDLE BEARINGS   BALL BEARINGS

Fig. 5-4 Types of antifriction bearings

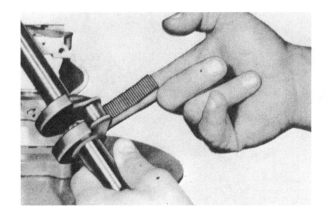

Fig. 5-5 Needle bearings

QUALITY DESIGNATION OF OIL AND SAE NUMBER

Selecting engine lubricating oils can be confusing. One must be aware of different viscosities, different qualities, different refining companies, different additives, and the meaning of many advertising phrases. Generally, lubricating oils used for various small gasoline engines are the same ones that are used for automobile engines. However, two-stroke cycle engines almost always use special two-cycle oil.

Quality Designation by the American Petroleum Institute (API)

Figure 5-6 is an example of lubricating instructions given on the decal. The API classifications are usually found on the top of the oil can and refer to the quality of the oil.

Oils for Service SE (Service Extreme). SE oils are suitable for the most severe type of operation beginning with 1972 models and some 1971 automobiles. Extreme conditions of start-stop driving; short trip, cold weather driving; and high speed, long distance, hot weather driving can be handled by this oil. The oil meets the requirements of automobiles that are equipped with emission control devices and engines operating under manufacturers' warranties.

Oils for SD (Service Deluxe). Most small engine manufacturers approve of the use of SD (formerly MS) oil in their engines. These oils provide protection against high and low temperature engine deposits, rust, corrosion, and wear.

Oils for Service SC. This oil was also formerly classified as suitable for MS. It has much the same characteristics as SD but is not quite as effective as SD oil. SC oil was developed for auto engines of 1964-67, while SD oil was developed for engines of 1968-70 manufacture.

Oils for Service SB (Formerly MM). This oil is recommended for moderate operating conditions such as moderate speeds in warm weather; short-distance, high-speed driving; and alternate long and short trips in cool weather. It is satisfactory for certain older autos but not for new autos under warranty. It is seldom recommended for small engine use.

Oils for Service SA (Formerly ML). No performance requirements are set for this oil. It is straight mineral oil and may be suitable for some light service requirements. SA oil is not recommended by small engine manufacturers.

The new classifications for oils suitable for diesel service are CA, CB, CC, and CD. These replace the old classifications of DG, AM (Supp-1), DM (MIL-L2104B), and DS respectively.

It is wise to buy the best quality oil for an engine. Sacrificing on quality for a few cents savings may be more expensive in the long run due to engine wear.

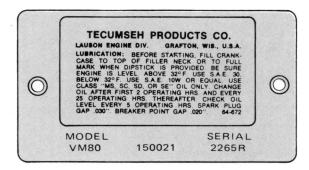

Fig. 5-6 Lubricating instructions on decal

Viscosity Classification by the
Society of Automotive Engineers

The SAE (Society of Automotive Engineers) number of an oil indicates its *viscosity,* or thickness. Oils may be very thin (light), or they may be quite thick (heavy). The range is from SAE 5W to SAE 50; the higher the number, the thicker the oil. The owner should consult the engine manufacturer's instruction book for the correct oil. SAE 10W, SAE 20W, SAE 20, and SAE 30 are the most commonly used oil weights. Usually, manufacturers recommend SAE 30 for summer use and either SAE 10W or SAE 20W for cold, subfreezing weather. In winter, thinner oil should be added since oil thickens in cold weather. The *W,* as in 5W, 10W, and 20W, indicates that the oil is designed for service in subfreezing weather.

Multigrade Oils. Multigrade oils may also be used in small gasoline engines, figure 5-7. These oils span several SAE classifications because they have a very high viscosity index. Typical classifications are SAE 5W-20, or 5W-10W-20; 5W-30 or 5W-10W-20W-30; SAE 10W-30 or 10W-20W-30; 10W-40, and 20W-40. The first number represents the low temperature viscosity of the oil; the last number represents its high temperature viscosity. For example, 10W-30 passes the viscosity test of SAE 10W at low temperatures and the viscosity test of SAE 30 at high temperatures. These oils are also referred to as all-season, all-weather oils, multiviscosity, or multiviscosity grade oils.

Motor Oil Additives

The best oil cannot do its job properly in a modern engine if additives are not blended into the base oil.

Pour-Point Depressants. Pour-Point depressants keep the oil liquid even at very low temperatures when the wax in oil would otherwise thicken into a buttery consistency making the oil ineffective.

Oxidation and Bearing Corrosion Inhibitors. These additives prevent the rapid oxidation of the oil by excessive heat. Viscous, gummy materials are formed without these inhibitors. Some of these oxidation products attack metals such as lead, cadmium, and silver which are often used in bearings. The inhibiting compounds are gradually used up and, thus, regular oil changes are needed.

Rust and Corrosion Inhibitors. These inhibitors protect against the damage that might be caused by acids and water which are by-products of combustion. Basically, acids are neutralized by alkaline materials much the same as vinegar can be neutralized with baking

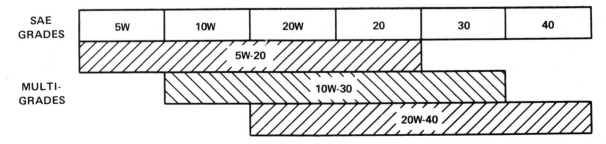

Fig. 5-7 Multigrade motor oils span several SAE (single) grades

soda. Special chemicals surround or capture molecules of water preventing their contact with the metal. Other chemicals which have a great attraction for metal form an unbroken film on the metal parts. These inhibitors are also used up in time.

Detergent/Dispersant Additives. This type of additive prevents the formation of sludge and varnish. Detergents work much the same as household detergents in that they have the ability to disperse and suspend combustion contaminants in the oil but do not affect the lubricating quality of the oil. When the oil is changed, all the contaminants are discarded with the oil so that the engine is kept clean. Larger particles of foreign matter in the oil either settle out in the engine base or are trapped in the oil filter if one is used.

Foam Inhibitors. Foam inhibitors are present in all high-quality motor oils to prevent the oil from being whipped into a froth or foam. The action in the crankcase tends to bring air into the oil; foaming oil is not an effective lubricant. Foam inhibitors called silicones have the ability to break down the tiny air bubbles and cause the foam to collapse.

LUBRICATION OF FOUR-CYCLE ENGINES

Since all moving parts of the engine must be lubricated to avoid engine failure, a constant supply of oil must be provided. Each engine, therefore, carries its own reservoir of oil in its crankcase where the main engine parts are located.

Basically, the oil is either pumped to or splashed on the parts and bearing surfaces that need lubrication. There are several lubrication systems used on small engines.

The following systems are among the most common.

- Simple Splash
- Constant-level Splash
- Ejection Pump
- Barrel-type Pump
- Full-pressure Lubrication

Simple Splash System

The *simple splash system* is perhaps the simplest system for lubrication. It consists of a splasher or dipper that is fastened to the connecting rod cap. There are several different sizes and shapes of dippers, figure 5-8. The dipper may be bolted on with the connecting rod cap or may be cast as a part of the connecting rod cap. Each time the piston nears the bottom of its stroke the dipper splashes into the oil reservoir in the crankcase, splashing oil onto all parts inside the crankcase. Since the engine is operating at 2000 to 3000 rpm, the parts are literally drenched by millions of oil droplets. Vertical crankshaft engines use an oil slinger that is driven by the

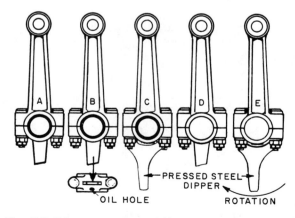

Fig. 5-8 Dippers used on different connecting rods

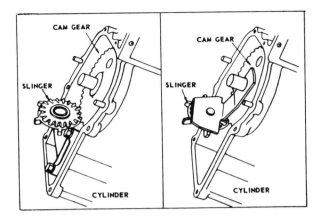

Fig. 5-9 Oil slinger

camshaft, figure 5-9. The slinger performs a similar function to the dipper.

Constant-level Splash System

The *constant-level splash system* has three refinements over the simple splash system: a pump, a splash trough, and a strainer, figure 5-10. With this system a cam-operated pump brings oil from the bottom of the crankcase into a splash trough. Again the splasher on the connecting rod dips into the oil splashing it on all parts inside the crankcase. The strainer prevents any large pieces of foreign matter from recirculating through the system. The pump maintains a constant oil supply in the trough regardless of the oil level in the crankcase.

Ejection Pump System

Ejection pumps of various types are found on many small engines, figure 5-11. With this method, a cam-operated or electric-operated pump draws oil from the bottom of the crankcase and sprays or squirts it onto the connecting rod. Some of the oil enters the connecting rod bearing through small holes; the remainder of the oil is deflected onto the other parts within the crankcase.

The parts of the ejection pump are the base, screen, check valve, spring, and plunger. The spring tends to push the plunger up but the plunger is driven down every revolution by the cam. A check valve allows the chamber to be filled with oil when the plunger goes up but when the cam pushes the plunger down,

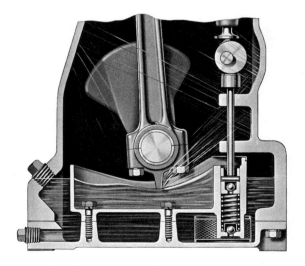

Fig. 5-10 Constant-level splash system used on some engines

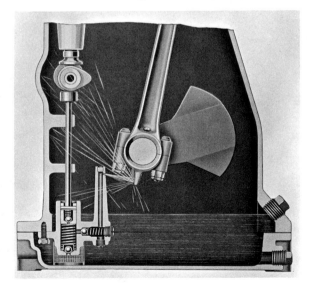

Fig. 5-11 Ejection pump as it is used on some engines

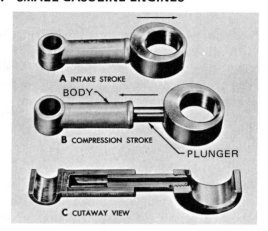

Fig. 5-12 Barrel-type pump

the check valve closes, and the trapped oil is squirted or sprayed from the pump.

Barrel-type Pump System

The *barrel-type pump* is also driven by an eccentric on the camshaft, figures 5-12 and 5-13. The camshaft is hollow and extends to the pump of the vertical crankshaft. As the

pump plunger is pulled out on intake, an intake port in the camshaft lines up allowing the pump body to fill. When the plunger is forced into the pump body on discharge, the discharge ports in the camshaft line up allowing the oil to be forced to the main bearing and to the crankshaft connecting rod journal. Small drilled passages are used to channel the oil. Oil is also splashed onto other crankcase parts.

Full-pressure Lubrication System

Full-pressure lubrication is found on many engines especially the larger of the small engines and on automobile engines. Lubricating oil is pumped to all main, connecting, and camshaft bearings through small passages drilled in these engine parts, figure 5-14. Oil is also delivered to tappets, timing gears, etcetera under pressure. The pump used is usually a positive displacement gear type. It is also common to use a splash system in conjunction with full-pressure systems.

LUBRICATING CYLINDER WALLS

Oil is splashed or sprayed onto cylinder walls and the piston rings spread the oil

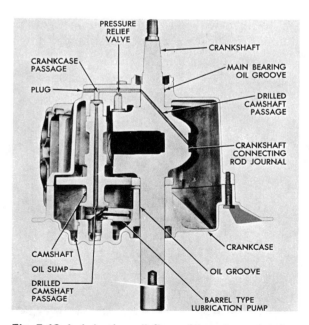

Fig. 5-13 Lubrication oil flow of barrel-type lubrication system

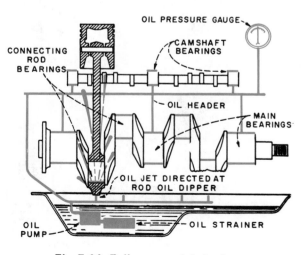

Fig. 5-14 Full-pressure lubrication

evenly for proper lubrication. Piston rings must function properly to avoid excessive oil consumption. The rings must put an even pressure on the cylinder walls and give a good seal. If piston ring grooves are too large, the piston rings begin a pumping action as the piston moves up and down. As a result, excess oil is brought into the combustion chamber. Worn cylinder walls, worn pistons, and worn rings can all add to high oil consumption as well as loss of compression and eventual loss of power.

BLOW-BY

Blow-by, the escape of combustion gases from the combustion chamber to the crankcase, occurs when piston rings are worn or are too loose in their grooves. Carbon and soot from burned fuel are forced into the crankcase by the rings. Much of the carbon is deposited around the rings which hinders their operation further.

Another damaging result of worn rings can be crankcase dilution. If raw, unburned gasoline collects in the combustion chamber, it can leak by the piston rings and into the crankcase. This dilutes the crankcase oil and reduces the oil's lubricating properties.

CRANKCASE BREATHERS

Four-stroke cycle engines do not have completely airtight, oiltight crankcases. Engine crankcases must breathe. Without the ability to breathe, pressures build up in the crankcase and cause oil seals to rupture or allow contaminants to remain in the crankcase. Pressure buildup may be caused by the expansion of the air as the engine heats up, by the action of the piston coming down the cylinder, and by the blow-by of combustion gases along cylinder walls.

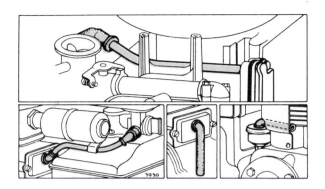

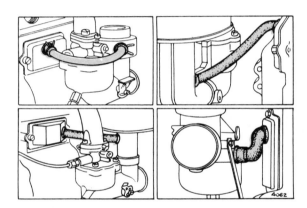

Fig. 5-15 Breather assemblies

Most single-cylinder engines use breathers, figure 5-15. These breathers allow air to leave but not reenter a reed-type check valve or a ball-type check valve. The breathers place the crankcase under a slight vacuum and are called closed breathers.

Open breathers allow the engine to breathe in and out freely and are usually equipped with air filters. They are located where the splashing of oil is not a problem. Open breathers are often incorporated with the valve access cover. If an engine with this type of breather is tipped on its side, oil may run out through it. Some breathers are vented to the atmosphere while others are vented back through the carburetor.

OIL CHANGES

The crankcase oil should be changed periodically. The exact number of engine operating hours between oil changes varies depending on the manufacturer. It may be as short a time as every 20 hours or as long a time as every 100 hours. (Every 25 hours is most common.) To ensure minimum engine wear and maximum engine life, follow the manufacturer's lubrication suggestions.

It is best to drain the oil when the engine is hot since more dirt and slightly more oil can be removed. Dirty oil should be replaced because it cannot give proper, high-quality lubrication.

LUBRICATION OF TWO-CYCLE ENGINES

Lubrication of the two-cycle engine is quite different from the four-cycle engine. Since the fuel mixture must travel through the crankcase, a reservoir of oil cannot be stored there. The lubricating oil is mixed with gasoline and then put into the gas tank. The lubricating oil for all crankcase parts enters the crankcase as a part of the fuel mixture. Millions of tiny oil droplets suspended in the mixture of gasoline and air settle onto the moving parts in the crankcase providing lubrication. Oil droplets are relatively large and heavy and quickly drop out of suspension. Much of the oil is carried on into the combustion chamber where it is burned along with the gasoline and air.

On a two-cycle engine, oil must be mixed with the gasoline, figure 5-16. The engine is not operated on gasoline alone. If it were, the heat of friction would burn up the engine in a short time. Some two-cycle engine manufacturers are developing and marketing engines with oil metering devices that eliminate the

need to premix the oil and gasoline. Mixing is done automatically in the correct proportions.

Too much oil mixed with the gasoline results in incomplete combustion and a very heavy exhaust. Deposits build up rapidly in the engine, possibly fouling the tip of the spark plug and collecting around the exhaust ports. Clogged exhaust ports reduce the engine's power output considerably and can cause overheating.

Too little oil mixed with the gasoline can cause the engine to overheat. It may even cause scuffing and scoring of the cylinder walls, bearing damage, or the piston to seize in the cylinder.

In preparing the mixture of gasoline and oil for most two-cycle engines, observe the following rules:

- Mix a good grade of regular gasoline (leaded or unleaded) and oil in a separate container. Do not mix in the gas tank unless it is a remote tank such as is found on many outboard motors.

Fig. 5-16 Thorough mixing of oil and gasoline is necessary for two-cycle engines.

Fig. 5-17 Proper mixing of oil and gasoline for this two-cycle portable generator is essential.

- Pour the oil into the gasoline to ensure good mixing, and shake the container well. If poorly mixed, the oil settles to the bottom of the tank causing difficult starting.

- Strain the fuel with a fine mesh strainer when pouring it into the tank to prevent any moisture from entering the tank.

- Use the oil that is specified for the particular two-cycle engine. Most manufacturers specify their own brand of two-cycle oil. In an emergency other oils may be used, usually SAE 30SB (formerly MM) or SAE 30SD (formerly MS) nondetergent oil. The manufacturer's brand, however, provides the best lubrication with a minimum of deposit formation.

- Mix gasoline and oil in the proportions recommended by the engine manufacturer. One common proportion is three-fourths pint of oil to one gallon of gasoline when breaking in a new engine,

Fig. 5-18 Most outboard motors use two-cycle engines which require mixing of gasoline and oil.

and one-half pint of oil to one gallon of gasoline for normal use.

NOTE: Mixing proportions vary widely; check the engine's instruction book.

MIXING RATIOS					
OUNCES TO BE MIXED WITH GASOLINE					

For a Dilution Ratio of	Gallons of gasoline					
	1	2	3	4	5	6
	Ounces of Two-Cycle Oil					
24:1	5	11	16	21	27	32
20:1	6	13	19	26	32	38
16:1	8	16	24	32	40	48

Add oil to gasoline in proportions recommended by manufacturer.

ADDITIONAL LUBRICATION POINTS

Whether the engine has two-cycle or four-cycle lubrication, there may be other lubrication to consider besides the crankcase area. On an outboard engine do not neglect the lower unit which needs a special gear lubricant at several points, figure 5-19. If an engine is powering an implement or other machinery, there may be transmissions, gear boxes, chains, axles, wheels, shafts, and linkages that need periodic lubrication. Lubricate these additional parts with the oil or lubricant recommended by the manufacturer.

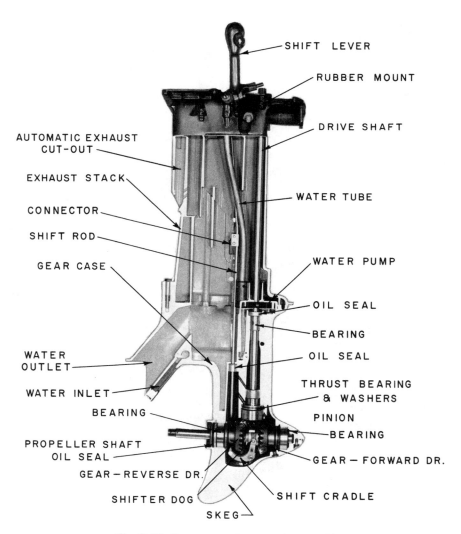

SHIFT LEVER

RUBBER MOUNT

DRIVE SHAFT

AUTOMATIC EXHAUST CUT-OUT

EXHAUST STACK

CONNECTOR

SHIFT ROD

GEAR CASE

WATER TUBE

WATER PUMP

OIL SEAL

BEARING

OIL SEAL

WATER OUTLET

WATER INLET

BEARING

PROPELLER SHAFT OIL SEAL

GEAR – REVERSE DR.

SHIFTER DOG

SKEG

THRUST BEARING & WASHERS

PINION

BEARING

GEAR – FORWARD DR.

SHIFT CRADLE

Fig. 5-19 Gear case — lower unit assembly

REVIEW QUESTIONS

1. What is the main job of a lubricant?
2. Explain how the lubricating oil also:
   a. Seals power
   b. Helps to dissipate heat
   c. Keeps the engine clean
   d. Cushions bearing loads
   e. Protects against rusting
3. What are the three types of friction bearings?
4. What is an antifriction bearing?
5. What does an oil's SAE number refer to?
6. What is a multigrade oil?
7. Explain how a detergent oil works.
8. What are the quality designations of oil?
9. List five common four-cycle engine lubrication systems.
10. How are cylinder walls lubricated?
11. Explain blow-by.
12. Briefly explain how two-cycle lubrication is accomplished?
13. What is one common proportion of oil to gasoline for two-cycle engines?
14. What kind of oil should be used for two-cycle engines?

CLASS DISCUSSION TOPICS

- Discuss how friction causes heat and wear.
- Discuss the jobs a lubricant performs.
- Discuss the types of bearings used on a demonstration engine.
- Discuss the good features of the various lubrication systems; also discuss the system's weak points, if any.

CLASS DEMONSTRATION TOPICS

- Illustrate the theory of lubrication using round pencils and a book.
- Demonstrate oil viscosity by showing and pouring several different weights of oil.
- Remove and inspect oil control rings.

- Remove an ejection pump from an engine and demonstrate the pump's operation.

- Remove and inspect the dipper on a splash lubrication system.

- Examine a crankshaft and point out the various bearing surfaces.

- Examine several cans of lubricating oil and determine the SAE number and quality designation.

- Demonstrate the correct mixing procedure for gasoline and oil for two-cycle engines.

LABORATORY EXPERIENCE 5-1
DRAIN OIL AND REFILL CRANKCASE

The crankcase oil should be changed periodically. The exact number of engine operating hours between oil changes varies from manufacturer to manufacturer. It may be as short a time as every 20 hours or as long a time as every 100 hours. (Every 25 hours is most common.) True, the manufacturer's suggested oil change interval can be stretched, but to insure minimum engine wear and maximum engine life follow the manufacturer's lubrication suggestions.

Manufacturers recommend high quality oil such as SE, SD, or SC. Avoid the use of light duty oil. The viscosity recommended is often SAE 30 for summer and SAE 20W for winter.

It is a good idea to drain the oil when the engine is hot since more dirt and slightly more oil can be removed. Remember, dirty oil should be replaced because it will not give proper high-quality lubrication.

STUDENT ASSIGNMENT

You are to drain the crankcase oil and then refill the crankcase with the correct type oil. Take care to prevent a mess; a good workman is not sloppy. It is essential that you record your work in a work record box as you complete each step of the procedure.

Procedure

Instructors may want to supplement or revise specific steps of this procedure since there are many types of engines. The following procedure is a general guide.

1. Remove the add-oil plug.
2. Loosen and carefully remove the drain plug. Do not drop the plug when it comes out of the engine. Be sure to have a container directly under the drain hole.
3. Allow the oil to drain, then tip the engine slightly to get the last bit of oil from the engine.
4. Replace the drain plug.
5. Refill the crankcase with clean oil of the correct quality and viscosity.
6. Check the oil level with the dipstick if the engine has one. On engines not having a dipstick, fill the crankcase until the oil can be seen at the hole or in the fill pipe.
7. Replace add-oil plug.

WORK RECORD BOX

PART	DISASSEMBLY (nuts, bolts, etc.)	OPERATION PERFORMED	TOOL USED

REVIEW QUESTIONS

After draining and refilling the crankcase, answer the following questions.

1. On most engines, how often should lubricating oil be changed?
2. What quality oil is generally used?
3. What viscosity (SAE No.) oil is often used in the summer? Winter?
4. Why is it a good idea to drain the oil just after the engine has been operated and it is hot?

LABORATORY EXPERIENCE 5-2
EJECTION OIL PUMP (FOUR-CYCLE)

The oil pump must provide oil for all bearing surfaces within the crankcase. At any point in the engine where one part is moving against another, oil is needed to reduce friction and wear. Generally, the ejection pump sprays or squirts oil onto all parts in the crankcase. A similar pump action can be seen in the constant level splash system where oil is pumped into the splash trough.

The parts of the ejection pump are the base, screen, check valve, spring and plunger. The spring tends to push the plunger up but the plunger is driven down every revolution by a cam (usually on the camshaft). A check valve allows the chamber to be filled with oil when the plunger goes up. When the cam pushes the plunger down, the check valve closes and the trapped oil is squirted or sprayed from the pump.

NOTE: Any engine with a camshaft driven pump that sprays or squirts oil onto the parts or pumps oil into the splash trough is suitable for this laboratory experience.

STUDENT ASSIGNMENT

You are to disassemble the engine exposing the pump and then remove the pump from the engine. Study the construction of the pump and then observe its operation, working it slowly by hand. It is essential that you record your work in a work record box as you complete each step of the disassembly procedure.

Disassembly Procedure (Ejection Pump)

Instructors may want to supplement or revise specific steps of this procedure since there are many makes of engines. The following procedure is a general guide.

1. Remove the air shrouding if necessary.
2. Drain the fuel system and oil bath air cleaner.
3. Drain the oil from the crankcase.
4. Remove the crankcase from the base. Note that there is a gasket between these sections.
5. Looking up into the bottom of the crankcase, locate the crankshaft, camshaft, and oil pump.
6. Remove the oil pump from the engine.
7. Disassemble the pump and study its main parts.

WORK RECORD BOX

PART	DISASSEMBLY (nuts, bolts, etc.)	OPERATION PERFORMED	TOOL USED

8. Reassemble the pump.
9. Study pump operation by:
   a. Pouring the oil into the engine base.
   b. Placing the oil pump base in the oil supply.
   c. Using a finger on the top of the plunger as a substitute for cam action, work the pump.

> CAUTION: Have a shield to stop the oil if it comes out of the pump with too much force.

Reassembly Procedure

Reverse the disassembly procedure. A new gasket may be needed between the base and crankcase.

REVIEW QUESTIONS

Study the ejection oil pump and its operation and then answer the following questions.

1. What are the main parts of the pump?
2. Is the oil discharge directed at any particular part within the crankcase or just the crankcase in general?
3. Why is a gasket necessary between the base and crankcase?
4. If the engine is operating at 3600 rpm, how many pump strokes are there each minute?
5. Can a drop in the level of crankcase oil affect the pump's output? Explain.

LABORATORY EXPERIENCE 5-3
SIMPLE SPLASH OIL SYSTEM
(HORIZONTAL CRANKSHAFT)

The simple splash oil system is an uncomplicated but very effective lubrication system. It consists of a dipper that is fastened to the connecting rod cap. At each revolution the dipper sweeps down and through the oil reservoir in the crankcase. Oil is splashed onto all parts that are within the crankcase. There is a drenching spray of oil in the crankcase when the engine is operating.

The dipper may be of several different sizes and shapes. The dipper may be bolted on with the connecting rod cap or it may be cast as a part of the connecting rod cap.

STUDENT ASSIGNMENT

You are to disassemble the engine, exposing the oil dipper. Slowly revolve the crankshaft and study the path of the dipper. Remove the dipper and study its construction. It is essential that you record your work in a work record box as you complete each step of the disassembly procedure.

Disassembly Procedure

Instructors may want to supplement or revise specific steps of this procedure since there are many makes of engines. The following procedure is a general guide.

1. Remove the air shrouding if necessary.

2. Drain the fuel system and oil bath air cleaner.

3. Drain the oil from the crankcase.

4. Remove the crankcase from the base (on some engines remove the cover assembly from the crankcase). Note that there is a gasket between these sections.

5. Locate the oil dipper. Slowly revolve the crankshaft and observe the path of the dipper.

6. Remove the connecting rod cap and dipper assembly. Study the part.

WORK RECORD BOX

PART	DISASSEMBLY (nuts, bolts, etc.)	OPERATION PERFORMED	TOOL USED

Reassembly Procedure

Reverse the disassembly procedure. Use care to replace the dipper and connecting rod cap exactly as they came off. A new gasket may be needed between the base and the crankcase.

REVIEW QUESTIONS

Study the splash oil system and then answer the following questions.

1. Is the dipper on the engine case part of the connecting rod cap or is it a separate part?

2. If a one-cylinder engine is operating at 4000 rpm, how many times does the dipper splash into the oil reservoir per minute?

3. Can a drop in the level of crankcase oil affect the lubrication within the crankcase? Explain.

UNIT 6

Cooling Systems

OBJECTIVES   After completing this unit, the student should be able to:
- Discuss the construction of the water jacket and trace the path of cooling water through the engine.
- Explain the construction and operation of the water pump and trace the path of water through the pump.
- List the basic parts of the air-cooling system and trace the path of cooling air through the engine.

In internal combustion engines, the temperature of combustion often reaches over 4000° Fahrenheit, a temperature well beyond the melting point of the engine parts. This intense heat cannot be allowed to build up. A carefully engineered cooling system is, therefore, a part of every engine. A cooling system must maintain a good engine operating temperature with allowing destructive heat to build up and cause engine part failure, figure 6-1.

The cooling system does not have to dispose of all the heat produced by combustion. A good portion of the heat energy is converted into mechanical energy by the engine, benefiting engine efficiency. Some heat is lost in the form of hot exhaust gases. The cooling system must dissipate about one-third of the heat energy caused by combustion.

Engines are either air-cooled or water-cooled; both systems are in common use. Generally, air-cooled engines are used to power machinery, lawn mowers, garden tractors, chain saws, etc. The air-cooled system is usually lighter in weight and simpler; hence, its popularity for portable equipment, figure 6-2.

The water-cooled engine is often used for permanent installation or stationary power plants. Most automobile engines and outboard motors are water-cooled.

107

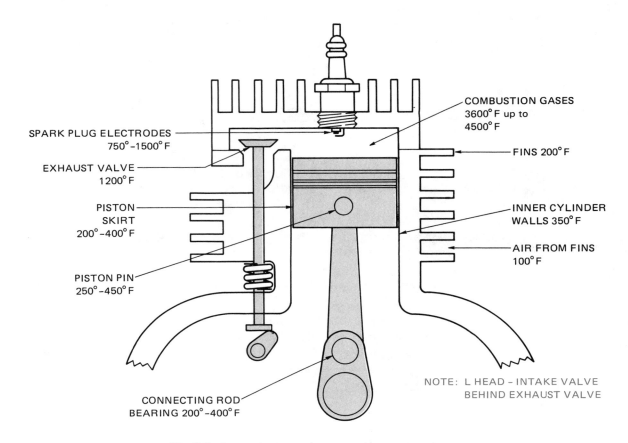

SPARK PLUG ELECTRODES
750°-1500° F

EXHAUST VALVE
1200° F

PISTON
SKIRT
200°-400° F

PISTON PIN
250°-450° F

CONNECTING ROD
BEARING 200°-400° F

COMBUSTION GASES
3600° F up to
4500° F

FINS 200° F

INNER CYLINDER
WALLS 350° F

AIR FROM FINS
100° F

NOTE: L HEAD – INTAKE VALVE
BEHIND EXHAUST VALVE

Fig. 6-1 Approximate engine operating temperatures

Fig. 6-2 Dirt bikes use air-cooled engines as they are lighter in weight than water-cooled engines.

AIR-COOLING SYSTEM

The air-cooling system consists of heat radiating fins, flywheel blower, and shrouds for channeling the air. The path of airflow can be seen in figure 6-3.

Heat radiating fins are located on the cylinder head and cylinder because the greatest concentration of heat is in this area. The fins increase the heat radiating surface of these parts allowing the heat to be carried away more quickly, figure 6-4.

The flywheel blower consists of air vanes cast as a part of the flywheel. As the flywheel revolves, these vanes blow cool air across the

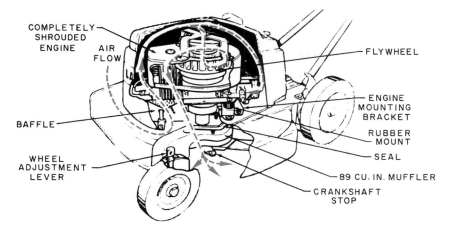

Fig. 6-3 The path of airflow on certain lawn mower engines

STRAIGHT FIN HEADS

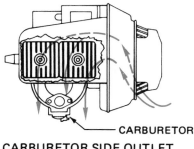

CARBURETOR SIDE OUTLET

fins carrying away the heated air and replacing it with cool air.

The shrouds direct the path of the cool air to the areas that demand cooling, figure 6-5. Shrouds must be in place if the cooling system is to operate at its maximum efficiency.

OBLIQUE FIN HEADS

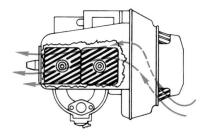

TOP OR END OUTLET

Fig. 6-4 Airflow across the cylinder head fins

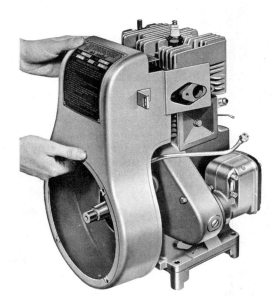

Fig. 6-5 The air shroud is an important part of the air-cooled engine.

Care of the Air-cooling System

The air-cooling system is almost trouble free. Several points, however, should be considered. Heat radiating fins are thin and often fragile, especially on aluminum engines. If through carelessness they are broken off, a part of the cooling system is gone. Besides losing some cooling capacity, hot spots can develop which may warp the damaged area. Also, it is easy for dust, dirt, grass, and oil to accumulate between the fins. Any buildup of foreign matter reduces the cooling system's efficiency. All parts of the system should be kept clean, especially the radiating fins.

The flywheel vanes should not be chipped or broken. Besides reducing the cooling capacity of the engine, such damage can destroy the balance of the flywheel. An unbalanced flywheel causes vibration and an excessive amount of wear on engine parts.

WATER-COOLING SYSTEM

Many small engines are water-cooled as are most larger engines. Such engines have an enclosed water jacket around the cylinder walls and cylinder head. Water jackets are relatively trouble free. However, damaging deposits can build up over a period of time. Salt corrosion, scale, lime, silt, etcetera, can restrict the flow of water and heat transfer. Water jackets are found on most outboard motors and on the majority of automobile engines. Cool water is circulated through this jacket picking up the heat and carrying it away.

Figure 6-6 shows a cross section of water passage in the head and block. Water cooling systems often found on stationary small gas engines or automobile engines include the following basic parts: radiator, fan, thermostat, water pump, hoses, and water jacket.

The water pump circulates the water throughout the entire cooling system. The hot water from the combustion area is carried from the engine proper to the radiator. In the radiator, many small tubes and radiating fins dissipate the heat into the atmosphere. A fan blows cooling air over the radiating fins. From the bottom of the radiator the cool water is returned to the engine.

Engines are designed to operate with a water temperature of between 160 degrees and 180 degrees Fahrenheit. To maintain the correct water temperature a thermostat is used in the cooling system. When the temperature is below the thermostat setting, the thermostat remains closed and the cooling water circulates only through the engine, figure 6-7A. However, as the heat builds up to the thermostat setting, the thermostat opens and the cooling water moves throughout the entire system, figure 6-7B.

Fig. 6-6 Cross section showing water passages in head and block

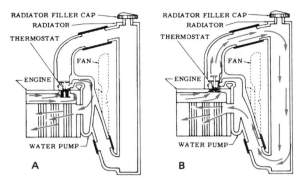

Fig. 6-7 (A) Thermostat closed: water recirculated through engine only. (B) Thermostat open: water circulated through both engine and radiator.

WATER COOLING THE OUTBOARD ENGINE

Cooling the outboard engine with water is a simpler process because there is an inexhaustible supply of cool water present where the engine operates. Outboards pump water from the source through the engine's water jacket and then discharge the water back into the source. A cutaway of an outboard motor is shown in figure 6-8.

The water pump on outboards is located in the lower unit. It is driven by the main driveshaft or the propeller shaft. The cool water is pumped up copper-tube passages to the water jacket. After the cooling water picks up heat, it is discharged into the exhaust area of the lower unit and out of the engine. Several types of water pumps are used on outboard motors. Plunger type pumps; eccentric rotor pumps, figure 6-9; and impeller pumps, figure 6-10, are some examples.

Many outboard motors, especially large horsepower models, are equipped with a thermostatically controlled cooling system, figure 6-11. The temperature of the water circulating through the water jacket is maintained at about 150° Fahrenheit.

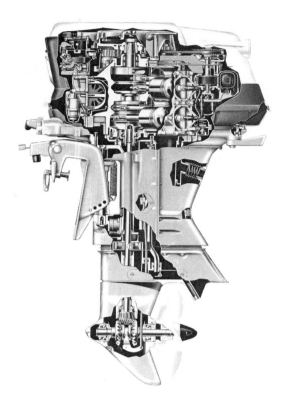

Fig. 6-8 Cutaway of an outboard motor

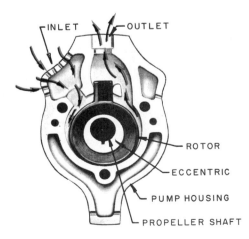

Fig. 6-9 An eccentric rotor water pump with the cover removed to show the eccentric and rotor

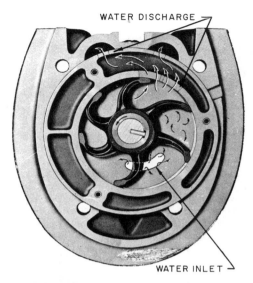

Fig. 6-10 An impeller water pump with the cover removed to show the impeller

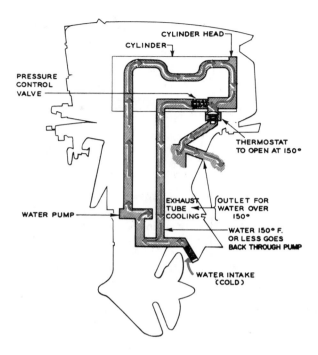

Fig. 6-11 A thermostatically controlled cooling system used on outboard engines

REVIEW QUESTIONS

1. How high can the temperature of combustion reach?

2. Does the cooling system remove all the heat of combustion? Explain.

3. Why are air-cooling systems often used for portable equipment?

4. What are the main parts of the air-cooling system?

5. Why are cylinders and cylinder heads equipped with fins instead of being cast with a smooth surface?

6. What are the basic parts of the water-cooling system found on stationary small gas engines?

7. What is the function of the thermostat?

8. What are several types of water pumps used on outboard motors?

CLASS DISCUSSION TOPICS

- Discuss how heat can damage engine parts.
- Discuss the path of heat flow from the inside of the engine to the outside.
- Discuss the advantages of the air-cooled engine.
- Discuss the advantages of the water-cooled engine.
- Discuss why it is best for the engine to operate at a constant temperature.

CLASS DEMONSTRATION TOPICS

- Trace the airflow through an air-cooled engine.
- Trace the water flow through an outboard engine.
- Disassemble various types of water pumps; show their operation and construction.
- Show how the air-cooling system can be damaged.

LABORATORY EXPERIENCE 6-1
WATER JACKET OF OUTBOARD MOTORS

The water jacket is constructed around the upper cylinder and head, the area of greatest heat concentration. Cooling water is pumped through the system picking up heat as it passes along. Water jackets are relatively trouble free. However, damaging deposits can build up over a period of time. Salt corrosion, scale, lime, silt, etcetera can restrict the flow of water and heat transfer. Water jackets are found on most outboard motors and on the vast majority of automobile engines.

STUDENT ASSIGNMENT

You are to disassemble and inspect the water jacket of an outboard engine. Determine the path of the cooling water within the jacket. Use a small horsepower outboard engine if possible. It is essential that you record your work in a work record box as you complete each step of the disassembly procedure.

Disassembly Procedure

Instructors may want to supplement or revise specific steps of this procedure since there are many makes of outboards. The following procedure is a general guide.

1. Remove the shrouds.

2. Remove the spark plugs.

WORK RECORD BOX

PART	DISASSEMBLY (nuts, bolts, etc.)	OPERATION PERFORMED	TOOL USED

3. Remove the machine screws in the top of the water jacket.

4. Remove the top of the water jacket. (The gasket between the engine block and the top of the water jacket is usually glued on and may be difficult to remove.)

5. Examine the construction of the water jacket. Note how deep it is and trace the path of the water flow.

Reassembly Procedure

Reverse the disassembly procedure. A new gasket must be installed if the engine is to be operated. All metal gasket surfaces must be absolutely clean before a new gasket is installed.

REVIEW QUESTIONS

Study the water jacket and then answer the following questions.

1. What can harm or damage the water jacket?

2. What would happen if an outboard engine were operated out of water?

3. What are the basic parts of the water-cooled engine?

LABORATORY EXPERIENCE 6-2
WATER PUMP OF OUTBOARD MOTORS

There are several different types of water pumps used on outboard motors. Among the more common are (1) eccentric rotor type, (2) impeller type, and (3) plunger type. The location of water pumps varies a great deal from manufacturer to manufacturer and, in fact, from model to model. However, the pump is always in the lower unit area. It is driven by the propeller shaft or the main drive shaft. The water intake may also be in a number of places on the underwater area of the lower unit. The water pump operates continually providing a supply of cool water for the engine.

STUDENT ASSIGNMENT

You are to disassemble and inspect the water pump of an outboard engine. There is such a wide variance in the location and disassembly procedure that it is not practical to include a general disassembly procedure in this assignment. Therefore, your instructor will give you instructions on just what parts to remove and how to do the work. It is essential that you record your work in a work record box as you complete each step of your disassembly.

WORK RECORD BOX

PART	DISASSEMBLY (nuts, bolts, etc.)	OPERATION PERFORMED	TOOL USED

REVIEW QUESTIONS

Study the water pump and answer the following questions.

1. What type of water pump does the engine have?
2. What drives the water pump?
3. Describe the location of the water intake.
4. Is the engine's cooling system thermostatically controlled?
5. Why is it important to check the engine to see if it is pumping water after starting it?

LABORATORY EXPERIENCE 6-3
AIR COOLING SYSTEM

The air-cooling system is simpler in construction and maintenance than the water-cooling system. No supply of water is necessary; the air around the engine is sufficient. The basic parts are the heat radiating fins, the flywheel vanes, and the air shroud.

The heat radiating fins around the upper cylinder provide a large radiating surface to liberate heat. The vanes on the flywheel provide a blast of cooling air. The air shroud channels the air from the flywheel to and across the radiating fins.

The main consideration is to keep the system parts clean, particularly the heat radiating fins. Dust, dirt, oil, grass clippings, etcetera can build up in a short time. Any buildup of foreign matter reduces the engine's cooling capacity. Actual damage or breakage of parts is possible through neglect or careless attempts at engine repair.

STUDENT ASSIGNMENT

You are to disassemble and inspect the basic parts of the air-cooling system. Inspect the parts for damage and cleanliness. It is essential that you record your work in a work record box as you complete each step of the disassembly procedure.

If you have proper ventilation and facilities for running the engine, you can demonstrate the air flow through the system before the disassembly and inspection portion of this assignment. Attach a thin, 5-inch strip of tissue (one that responds to light movement of air currents) to the end of a pencil. Start the engine and run it at normal operating speed. Bring the paper strip close to the various areas of the heat radiating fins and note the behavior of the paper. Then bring the paper close to the flywheel blower and note the movement of the strip. A flow of air currents should be clearly demonstrated. Be careful not to get too close to moving parts of the engine or to foul the system with bits of tissue.

Disassembly Procedure
1. Remove the air shroud, grass screen, recoil starter, etc.
2. Examine the air shroud for cleanliness and any possible damage.
3. Examine the flywheel air vanes for cleanliness and any possible damage.
4. Examine the heat radiating fins for cleanliness and any possible damage.

WORK RECORD BOX

PART	DISASSEMBLY (nuts, bolts, etc.)	OPERATION PERFORMED	TOOL USED

Reassembly Procedure

Reverse the disassembly procedure.

REVIEW QUESTIONS

Study the air-cooling system and answer the following questions.

1. Trace the airflow through the engine. Make a small sketch.
2. Explain the function of each of the three main parts of the air-cooling system.
3. Explain how to clean heat radiating fins.
4. How often should the system be cleaned?
5. How can the system be damaged?

UNIT 7

Ignition Systems

OBJECTIVES   After completing this unit, the student should be able to:
- Discuss electron theory and magnetism.
- Discuss how the basic parts of the magneto are constructed and how they are mounted on the engine.
- Explain the complete magneto cycle.
- Discuss spark advance.
- Discuss solid state ignition.
- Discuss battery ignition systems.

Small gasoline engines normally use a magneto for supplying the ignition spark. A *magneto* is a self-contained unit that produces the spark for ignition, figure 7-1. No outside source of electricity is needed to produce the spark for ignition. The magneto is a simple and very reliable ignition system. On small gas engines that do not have electric starters, lighting systems, radios, and other electrical accessories, a storage battery is not necessary. The magneto is, therefore, ideally suited for these small gasoline engines.

The basic parts of the magneto ignition system are the: permanent magnets, high-tension coil (primary and secondary), laminated iron core, breaker points, breaker cam, condenser, spark plug cable, and spark plug. Before discussing how these parts work together, a review is given of some essentials of electricity and magnetism which explains the construction and function of each individual part.

ELECTRON THEORY

All matter is composed of tiny particles called atoms. The atom is composed of electrons, protons, and neutrons, figure 7-2. The number and arrangement of these particles determines the type of atom: hydrogen,

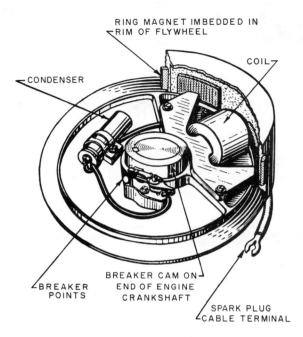

Fig. 7-1 Flywheel magneto

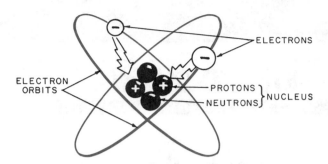

Fig. 7-2 Atomic structures: electron, proton, and neutron

oxygen, carbon, iron, lead, copper, or any other element. Weight, color, density, and all other characteristics of an element are determined by the structure of the atom. Electrons from an atom of copper are the same as electrons from any other element.

The electron is a very light particle that spins around the center of the atom. Electrons move in an orbit. The number of electrons orbiting around the center or nucleus of the atom varies from element to element. The electron has a negative (–) electrical charge.

The proton is a very large and heavy particle in relation to the electron. One or more protons form the center or nucleus of the atom. The proton has a positive (+) electrical charge.

The neutron consists of an electron and proton bound tightly together. Neutrons are located near the center of the atom. The neutron is electrically neutral; it has no electrical charge.

Atoms are normally electrically neutral; that is, the number of electrons and protons are the same, cancelling out each other's electrical force. Atoms stay together because, unlike electrical charges, they attract each other. The electrical force of the protons holds the electrons in their orbits. Like electrical charges repel each other, so negatively charged electrons do not collide with each other.

In most materials, it is very difficult, if not impossible, for electrons to leave their orbit around the atom. Materials of this type are called nonconductors of electricity or *insulators*. Some typical insulating materials are glass, mica, rubber and paper. Electricity cannot flow through these materials.

In order to have electric current, electrons must move from atom to atom. Insulators do not allow this electron movement. However, in many substances, an electron can jump out of its orbit and begin to orbit in an adjoining or nearby atom. Substances which permit this movement of electrons are called *conductors* of electricity. Some typical examples are copper, aluminum, and silver.

Electron flow in a conductor takes place when there is a difference in electrical potential and there is a complete circuit or path for electron flow, figure 7-3. In other words, when the source of electricity is short of electrons, it is positively charged. Since unlike charges attract each other, electrons (being negatively charged) move toward the positive source.

A source of electricity can be produced or seen in three basic forms: mechanical, chemical and static. Electricity is produced mechanically in the electrical generator which is commonly connected to water power or steam turbines. The electricity used in homes and factories is produced mechanically. In the magneto, mechanical energy is used to rotate the permanent magnet. Electricity produced by chemical action is seen in the storage battery and dry cell. Static electricity can be seen in nature when lightning strikes. The lightning occurs when the air insulation breaks down and electrons are in a positive area. The lightning may be between clouds, from cloud to earth, or from earth to cloud.

UNITS OF ELECTRICAL MEASUREMENT

There are three basic units of electrical measurement: *ampere* (rate of electron flow), *volt* (force or pressure causing electron flow), and *ohm* (resistance to electron flow).

Ampere

The *ampere* is the measurement of electrical current — the number of electrons flowing past a given point in a given length of time. If a person stands at a point on a wire and could count the electrons passing by in one second, 6,250,000,000,000,000,000 electrons would be counted, which equals one ampere of current. To help visualize amperage, think of water flowing in a pipe. A small pipe might deliver two gallons of water a minute. A larger pipe might deliver five gallons of water a minute. Electric wires are generally the same; larger wires can handle more amperage or electron flow than smaller wires.

Volt

The *volt* is the measurement of electrical pressure or the difference in electrical potential that causes electron flow in an electrical circuit. The energy source is short of electrons and the electrons in the circuit want to go to the source. The pressure to satisfy the source is called *voltage*. Voltage might be compared to the pressure that water in a high tank places on the pipe located at the street level. The higher the water pressure, the faster the water flow from a pipe below. Likewise, a higher voltage tends to cause greater flow of electrons.

Ohm

The *ohm* is the unit of electrical resistance. Every substance puts up some resistance to the movement of electrons. Insulators such as porcelain, oils, mica, and glass put up a great resistance to electron flow. Conductors such as copper, aluminum and silver put up very little resistance to electron flow. Even though conductors readily permit the flow of electric current they do tend to put up some resistance. In the water pipe example, this resistance can be seen as the surface drag by the sides of the pipe, or scale and rust in the pipe. Using a larger pipe or, electrically, a larger wire is one way of reducing resistance.

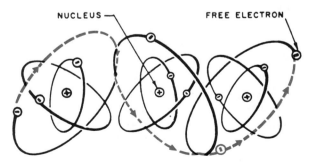

Fig. 7-3 Current: flow of electrons within a conductor

OHM'S LAW

In every example of electricity flowing in an electrical circuit, amperes, volts, and ohms each play their part; they are related to each other. This relationship is stated in *Ohm's Law*.

$$\text{Amperes (rate)} = \frac{\text{volts (potential)}}{\text{ohms (resistance)}}$$

The formula is usually abbreviated to:

$$I = \frac{E}{R}$$

For example, if the voltage is 6 and the resistance is 12 ohms, the current is calculated as follows.

$$I = \frac{E}{R} \quad I = \frac{6}{12} \quad I = 0.5 \text{ amperes}$$

The formula can be written to find the resistance, or the voltage.

$$R = \frac{E}{I} \quad E = IR$$

MAGNETISM

Most people have experimented with a magnet at one time, watching it pick up steel objects and attracting or repelling another magnet. These effects are not entirely explained but scientists generally agree on the molecular theory of magnetism. Molecules are the smallest divisions of substance that are still recognizable as that substance. Several different atoms may make up one molecule. For example, a molecule of iron oxide contains atoms of iron and oxygen. In many substances the atoms in the molecules are more positive at one spot and more negative at another spot. This is termed a *north pole* and a *south pole*. Usually the poles of adjoining molecules are arranged in a random pattern and there is no magnetic force since their

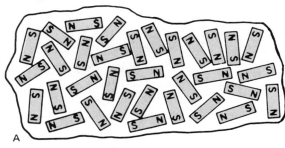

UNMAGNETIZED

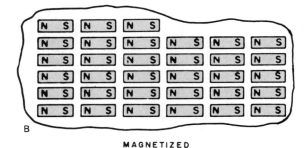

MAGNETIZED

Fig. 7-4 In an unmagnetized bar the molecules are in a random pattern. In a magnetized bar the molecules align their atomic poles.

effects cancel one another, figure 7-4A. However, in some substances such as iron, nickle, and cobalt the molecules are able to align themselves so that all north poles point in one direction and all south poles point in the opposite direction, figure 7-4B. The small magnetic forces of many tiny molecules combine to make a noticeable magnetic force. In magnets, like poles repel each other and unlike poles attract each other just as like and unlike electrical charges react, figure 7-5. Electricity and magnetism are very closely tied together.

Some substances keep their molecular alignment permanently and are classed as *permanent magnets*. Hard steel has this ability. Other materials such as a piece of soft iron (a nail), can attain the molecular alignment of a magnet only when it is in a magnetic field. As soon as soft iron is removed

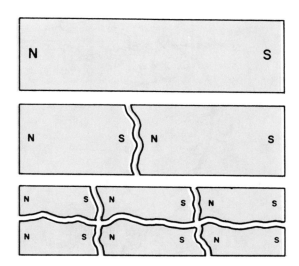

Fig. 7-5 A magnetic field surrounds every magnet. Like poles repel each other; unlike poles attract each other.

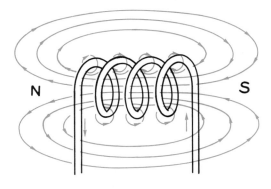

Fig. 7-6 Current flows in the wire as it moves down through the magnetic field.

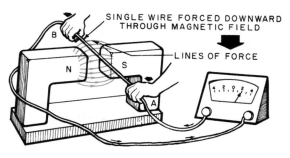

Fig. 7-7 When electricity flows through a coil of wire, a magnetic field is set up around the coil.

from the magnetic field, its molecules disarrange themselves and the magnetism is lost. Such substances are termed *temporary magnets.*

More than 100 years ago Michael Faraday discovered that magnetism could produce electricity. Magnetos used on gasoline engines apply this discovery. Faraday found that if a magnet is moved past a wire, electrical current starts through the wire. If the magnet is stopped near the wire, the current stops. Electricity flows only when the magnetic field or magnetic lines of force are being cut by the wire, figure 7-6.

Another principle that the magneto uses is that when electrons flow through a coil of wire, a magnetic field is set up around the coil, figure 7-7. The coil itself becomes a magnet. Therefore, when electrons flow through the coils in a magneto, a magnetic field is set up.

The principle of the transformer and induced voltage is also used in the magneto. In a transformer there is a primary coil and a secondary coil wound on top of the primary; the two are insulated from each other, figure 7-8. These coils are wound on a soft iron core. When alternating current passes through the primary coil there is an alternating magnetic field set up in the iron

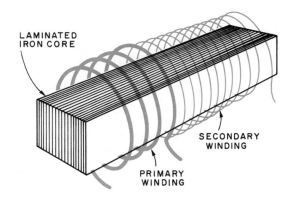

Fig. 7-8 A simple transformer

core. The magnetic lines of force cut the secondary coil and induce an alternating voltage within the coil. The voltage produced depends on the ratio of windings in the primary coil and secondary coil. If there are more windings in the secondary than in the primary, the secondary voltage is higher; a *step-up transformer*. If there are more windings in the primary than in the secondary, the secondary voltage is smaller; a *step-down transformer*. Although the magneto does not operate on alternating current, it does use the principle of the step-up transformer.

In a small gasoline engine magneto, there is a magnetic field which induces current in the primary coil, thus setting up a magnetic field around both the primary and secondary coils. At the point of maximum current, the circuit is broken in the primary. Electrons can no longer flow; therefore, the magnetic field collapses. This rapidly collapsing magnetic field induces a very high voltage igniting the fuel mixture.

Before studying in detail how the magneto operates, it is useful to understand the function of each part. The seven basic parts which work together to produce the ignition spark are the permanent magnets, high-tension coil, laminated iron core, breaker points, condenser, spark plug lead, and spark plug. An example of a magneto ignition system is shown in figure 7-9.

PERMANENT MAGNETS

Permanent magnets are usually made of an alloy called *alnico,* a combination of aluminum, nickle, and cobalt. This magnet is quite strong and keeps its magnetism for a very long time. On a flywheel magneto the magnet is cast into the flywheel and cannot be removed. The other magneto parts are often mounted on a fixed plate underneath the flywheel. The

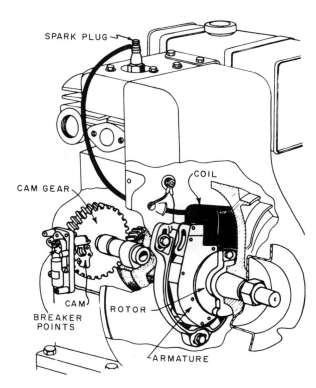

Fig. 7-9 The basic parts of a magneto ignition system

magnet revolves around the other parts of the magneto. Sometimes the other magneto parts are mounted near the outside rim of the flywheel and the permanent magnets pass by each revolution.

Rotor-type permanent magnets are also in use, figure 7-10. With this type of construction the permanent magnet rotor may be mounted on the end of the crankshaft or the rotor may be geared to the crankshaft. The

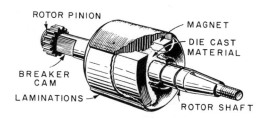

Fig. 7-10 Rotor-type permanent magneto

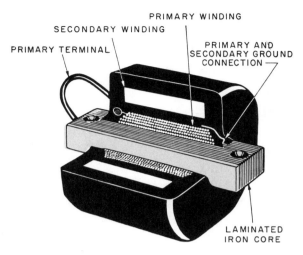

Fig. 7-11 Construction of a high-tension magneto ignition system

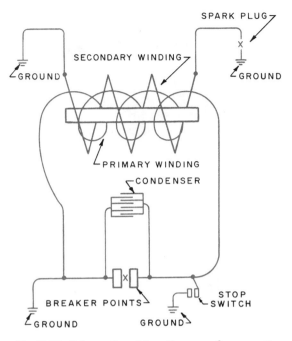

Fig. 7-12 Schematic wiring diagram of a magneto

rotor lies within the other magneto parts. Care should be taken not to drop or pound the magnet as this causes it to lose some of its magnetism.

HIGH-TENSION COIL (PRIMARY AND SECONDARY WINDINGS)

Refer to figure 7-11. The primary winding of wire consists of about 200 turns of a heavy wire (about 18 gauge) wrapped around a laminated iron core. The primary coil is in the electrical circuit containing the breaker points and condenser, figure 7-12. When a magnet is brought near this coil and iron core, magnetic lines of force cut the coil and electrical current is produced. Usually this current flows from the coil through the closed breaker points and is grounded. This makes a complete circuit.

The secondary winding of wire consists of about 20,000 turns of a very fine wire wrapped around the primary coil. The secondary coil is in the electrical circuit containing the spark plug. When current flows in the primary, a magnetic field gradually expands around both

coils. However, voltages produced in the secondary are quite small. When the breaker points open, the circuit is broken, electricity stops flowing and the magnetic field suddenly collapses. The suddenly collapsing magnetic field induces very high voltage in the secondary coil which is enough to jump the spark gap in the spark plug.

LAMINATED IRON CORE

The laminated iron core is made of many strips of soft iron fastened tightly together, figure 7-13. The soft iron core helps to strengthen the magnetic field around the primary and secondary coils, but the core cannot retain the magnetism and become permanently magnetized. The purpose of using many strips instead of a solid core is to reduce eddy currents which create heat in the core. The general shape of the laminated core

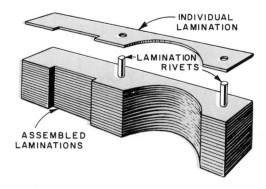

Fig. 7-13 Laminated iron core

may vary from magneto to magneto but its function remains the same.

BREAKER POINTS

The breaker points are made of tungsten and mounted on brackets, figure 7-14. The two points are usually closed or touching each other providing a path for electron flow. However, just before the spark is desired, the breaker points are opened by cam action breaking the electrical circuit. Breaker points open and close anywhere from 800 times per minute to 4500 or more times per minute depending on the engine speed.

When the breaker points are open they are usually separated by 0.020 of an inch; the

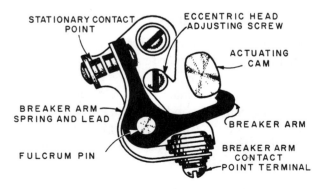

Fig. 7-14 Breaker point assembly

exact opening varies from magneto to magneto. This separation is critical to the function of the magneto. Therefore, the breaker point gap must be adjusted correctly.

BREAKER CAM

The breaker cam actuates the breaker points. One bracket of the breaker assembly rides on this cam. As the cam rotates, it opens and closes the breaker points.

On most two-cycle engines, this cam is mounted on the crankshaft and opens the normally closed breaker points once each revolution. On most four-cycle engines, the breaker cam is mounted to operate from the camshaft which is turning at one-half crankshaft speed. By mounting the breaker points here, they open once every two revolutions of the crankshaft. This provides a spark for the power stroke but none for the exhaust stroke. The shape of the cam depends on the number of cylinders and the design of the magneto, figure 7-15.

CONDENSER

The condenser acts as an electrical storage tank in the primary circuit, figures 7-16 and 7-17. When the breaker points open quickly, the electrons tend to keep flowing. If no condenser was present, a spark could jump across the breaker points. If this happened, the breaker points would soon burn up resulting in a weakening effect on the voltage produced by the magneto secondary coil. The condenser provides an electrical storage tank for this last surge of electron flow. The action of the condenser provides for a very abrupt interruption of the primary circuit. The condenser's discharge helps to induce high voltages in the secondary coil as the magnetic fields reverse. The condenser can be easily located because it usually looks like a miniature tin can.

SPARK PLUG CABLE

The spark plug cable connects the secondary coil and the spark plug providing a path for the high-tension voltage. This part is often referred to as the high-tension lead.

SPARK PLUG

The spark plug is a vital part of the ignition system. In this part the resulting work of the magneto parts is seen. Basically, the spark plug consists of a shell, ceramic insulator, center electrode, and ground electrode. The

| ONE LOBE CAM | TWO LOBE CAM | THREE LOBE CAM | FOUR LOBE CAM | SIX LOBE CAM |

Fig. 7-15 Common breaker cam shapes

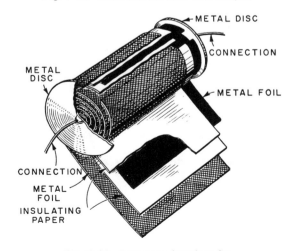

Fig. 7-16 Cutaway showing the
construction of a condenser

two electrodes are separated by a gap of about 0.035 of an inch. The path of electricity is down the center electrode and across the air gap to the ground electrode. The voltage must jump the air gap. When a high-tension voltage of about 20,000 volts is reached, a spark jumps between the electrodes. This spark ignites the fuel mixture within the combustion chamber.

THE COMPLETE MAGNETO CYCLE

Refer to figure 7-18 for illustrations of the steps in the magneto cycle. When the permanent magnet is far away from the high-tension coil, it has no effect on the coil. As the permanent magnet comes closer and closer, the primary coil feels the increasing magnetic field; the coil is being cut by magnetic lines of force. Therefore, electrons flow

Fig. 7-17 Typical condensers

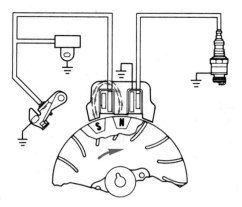

(A) Magnetic field as magnets approach the coil and laminated core

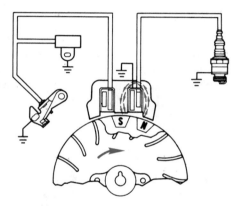

(B) Polarity of magnetic fields reverses rapidly

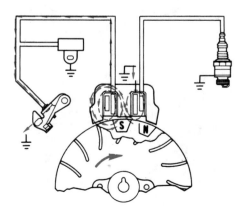

(C) Current flows in the primary circuit as magnetic fields reverse

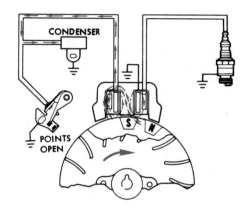

(E) Collapsing magnetic fields and condenser surge induces high voltage to jump the gap at the spark plug

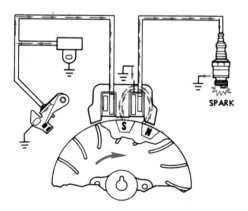

(D) Current stops as breaker points open

Fig. 7-18 The magneto cycle

in the primary coil. This current passes through the breaker points and into the ground. When the permanent magnet is nearly opposite the high-tension coil the magnetic field around both coils is reaching its peak. Also, the piston is reaching the top of its stroke compressing the fuel mixture.

At this time, the position of the permanent magnet causes the polarity of the laminated iron core and the coils to reverse direction. This change of direction is momentarily choked or held back by the coil. The breaker points then open interrupting the current flow and allowing the reversal and rapid collapse of the intense magnetic field that had been built up around the primary and secondary coils. Magnetic lines of force are cutting coils very rapidly and high voltages are induced in the coils. In the secondary coil, the voltage may reach 18,000 to 20,000 volts which is enough to jump across the spark gap in the spark plug. This spark ignites the fuel mixture and the piston is forced down the cylinder.

SPARK ADVANCE

When fuel burns in the combustion chamber, it does not explode and exert all of its power immediately. A very short period of time is required for the fuel to ignite and reach its full power. Even though this time lag is very short, it is very important in an engine operating at high speeds. For example, if the mixture was not ignited until the piston reached dead center, the piston would already be starting back down the cylinder before the full force of the burning fuel is reached. This results in loss of power. Therefore, it is necessary to ignite the fuel slightly before the piston reaches the top of its stroke to realize the full force of combustion. Causing the spark to occur earlier in the engine cycle is called *spark advance*.

The spark is advanced more and more as the engine speed increases because there is less time for combustion to take place. For example, in a four-cycle engine operating at 2000 rpm, each power stroke takes about 1/64 of a second. If the speed is increased to 4000 rpm, each power stroke takes only 1/128 of a second. If the speed is doubled, there is only one-half as much time for combustion to take place. Also, high speeds give the engine higher compression and a more explosive mixture. At high speeds the spark jumps the spark gap before the piston reaches the top of its stroke.

At slow speeds the spark is delayed and occurs later in the cycle, slightly before the piston reaches top dead center. At slow speeds there is more time for combustion to take place. Also, compression is lower and not as much explosive mixture is drawn into the combustion chamber.

Regulating spark advance is done by controlling the time that the breaker points open. Advancing the spark can be done automatically or manually. Both methods are commonly used on small gasoline engines.

MANUAL SPARK ADVANCE

Manual spark advance is usually accomplished by loosening the breaker point assembly and rotating it slightly to an advanced position. It is then locked in its new position.

Many outboard motors have a type of manual spark advance although it is generally referred to by manufacturers as a spark-gas synchronization system. The magneto plate on which the breaker points are mounted is located underneath the flywheel. This plate can be rotated through about 25 degrees. The breaker cam is securely mounted on the crankshaft; its relative position cannot be changed. Spark advance is obtained by moving the magneto plate so the breaker points are opened earlier in the cycle. The throttle and magneto plates are linked together so that opening the throttle also advances the spark. At full throttle the spark is fully advanced. At idling speeds the throttle is closed and the magneto plate is positioned for minimum spark advance.

AUTOMATIC SPARK ADVANCE

Automatic spark advance is usually accomplished by a centrifugal mechanism which is capable of changing the relative position of the breaker cam and the breaker points, figures 7-19 and 7-20. Here the breaker cam can be rotated through a small distance on its shaft, about 25 degrees. A spring holds the

cam in the retarded position at idling and slow speeds. As the engine speed increases, centrifugal force throws the mechanism's hinged weights outward. This outward motion overcomes the spring's tension and this motion is used to rotate the cam to a more advanced position. At full speed the cam has been rotated its full limit for maximum spark advance.

Some engines, especially larger types, also have a vacuum advance mechanism working with a centrifugal mechanism to provide more accurate spark advance, especially at slow speeds. Automatic spark advance can also be

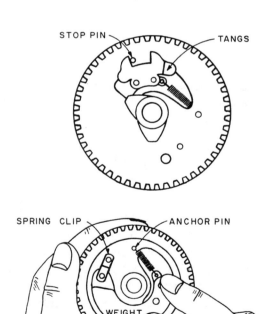

Fig. 7-19 Spark advance mechanism with breaker cam and weight mounted on the camshaft gear

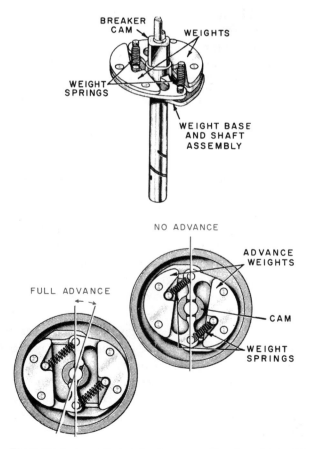

Fig. 7-20 Automatic spark advance mechanism changes the breaker cam's position in relation to the points

used to move the breaker point assembly to secure the proper advance.

IMPULSE COUPLING

When an engine is started by hand, it is turned over slowly. The voltages produced by the magneto depend upon the speed at which magnetic lines of force cut the primary coil. Therefore, the voltages produced for starting can be quite low, resulting in a weak spark.

Some engines are equipped with an impulse coupling device to supply higher voltages and a hotter spark for slow starting speeds. Using a tight spring and retractable pawls the rotation of the magneto's magnetic rotor can be stopped for all but the last few degrees of its revolution, figure 7-21. During the last few degrees of the revolution, the pawl retracts allowing the spring to snap the magnetic rotor past the firing position at a very high speed. A high voltage can be produced

because magnetic lines of force are cutting the primary coil rapidly.

MAGNETO IGNITION FOR MULTICYLINDER ENGINES

Magnetos are often used for multicylinder engines. However, having more than one cylinder to supply with a spark does present a problem. Two solutions to this problem are in common use today.

Two-, three- and four-cylinder engines can have magneto ignition by simply installing a separate magneto for each cylinder. This method is commonly used with outboard engines using a flywheel magneto. Instead of mounting just one magneto on the armature plate under the flywheel, a magneto is installed for each cylinder, figure 7-22. The breaker points of each magneto ride on the common cam. In two-cylinder engines the magnetos are 180 degrees apart; three-cylinder engines 120 degrees apart; and four-cylinder engines

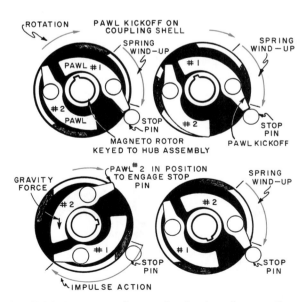

Fig. 7-21 Sequence of operation for impulse coupling for 180 degrees spark magneto

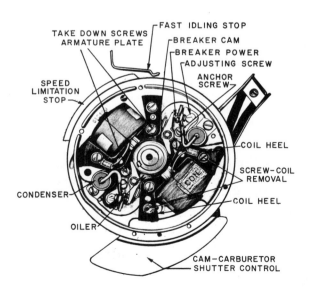

Fig. 7-22 Two complete magnetos are installed on this armature plate.

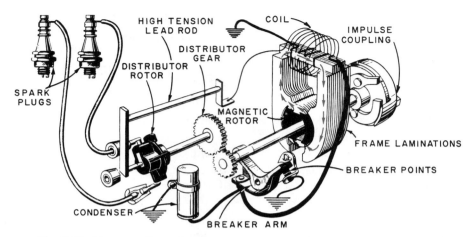

Fig. 7-23 Diagram of rotating-magnet magneto with jump-spark distributor

90 degrees apart. Each magneto functions separately to supply its cylinder with a spark.

Another solution is to use a distributor and a two-pole or four-pole magnetic rotor, figure 7-23. Only one complete magneto is used even though the engine may have two or four cylinders. In the case of a two-pole rotor used on a two-cylinder engine, the magneto can produce two sparks every revolution of the magnetic rotor, one every 180 degrees rotation. A two-lobed cam is used to open the breaker points twice each rotor revolution. The distributor rotor channels the high-tension voltage to the correct spark plug.

ENGINE TIMING

The spark must jump the spark gap at exactly the right time. This is just before the piston reaches top dead center. In many engines, the breaker cam is driven by the camshaft, therefore, the gear on the crankshaft and the camshaft gear must be assembled correctly. These two gears are usually marked with punch marks so the repairer can easily make the correct assembly, figure 7-24. Incorrect alignment of the gears can cause poor operation or no operation at all.

SOLID STATE IGNITION

Solid state ignition refers to the fact that solid state electronic parts (namely transistors), replace the breaker points. The breaker points in a conventional ignition system can be a source of trouble. The breaker cam is also eliminated so there are no moving parts in the system. Sometimes the systems are called *capacitor discharge systems, breakerless ignition, transistorized ignition* or *electronic ignition.* These systems can be used with just a conventional flywheel magnet as the source of the magnetic force, or they may be used

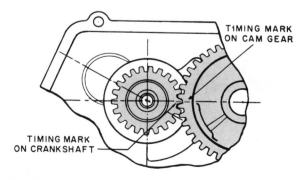

Fig. 7-24 Timing marks on crankshaft gear and cam gear must be aligned correctly.

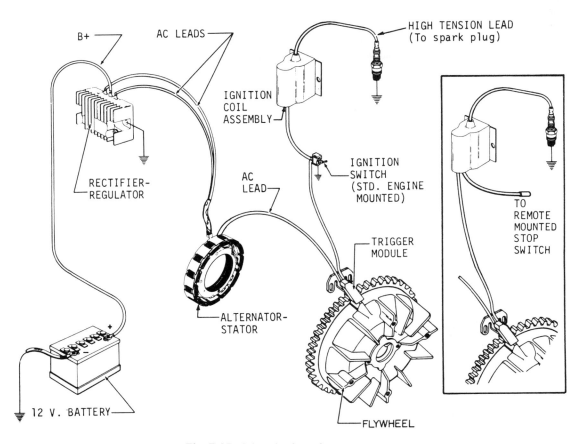

Fig. 7-25 A breakerless-alternator system

with a generator or an alternator battery system, figure 7-25.

The main components of the solid state ignition system are a generator or alternator coil, trigger module, ignition coil assembly, and flywheel magnets. The system has a conventional spark plug and spark plug lead. A schematic drawing of a solid state ignition system is shown in figure 7-26.

The trigger module contains transistor diodes which rectify the alternating current, changing it into direct current. Transistors control the flow of electric current by acting somewhat like a valve. Their resistance to flow can be changed by a small current to the transistors. In addition to the diode rectifiers, the trigger module contains a resistor, a sensing coil and magnet, and a silicon controlled rectifier (SCR). The silicon controlled rectifier acts as a switch.

The ignition coil contains primary and secondary windings similar to those of a conventional magneto coil plus a condenser or capacitor. Capacitors and condensers are basically the same electrically. The operation of the solid state ignition system follows this cycle:

1. The rotating magnet on the flywheel sets up an alternating current in the charging coil. This alternating current

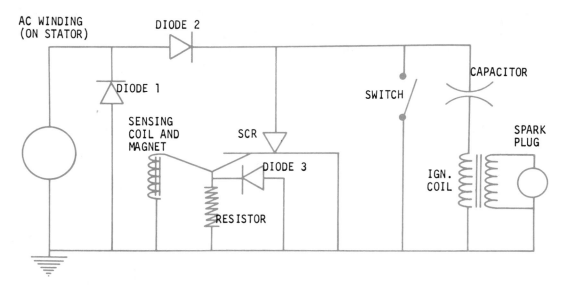

Fig. 7-26 Schematic of a solid state ignition system

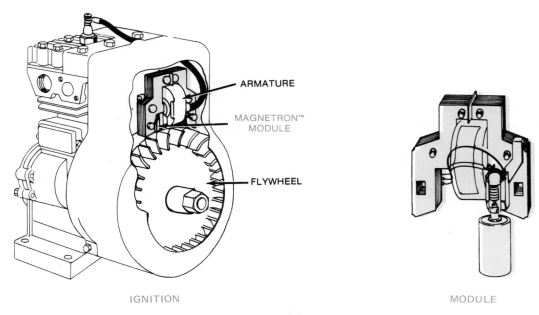

IGNITION

MODULE

Fig. 7-27 Solid state ignition "Magnetron^{TM}" used by Briggs and Stratton

is rectified into direct current and stored in the capacitor. The diode rectifiers permit flow in one direction only so the capacitor cannot discharge back through the diodes.

2. The magnet group passes the trigger coil, setting up a small current which triggers or gates the SCR. This gating makes the SCR conductive so that the stored voltage of the capacitor surges from the capacitor through the SCR. It then is applied across the primary winding of the ignition coil. This surge of energy sets up a magnetic field in the primary winding of the ignition coil. A magnetic field is also induced around the secondary coil, and voltages of about 30,000 are produced which is sufficient voltage to jump the spark gap at the spark plug.

Solid state ignition offers several advantages such as automatic retarding of the spark at starting speeds, longer spark plug life, faster voltage rise, and high-energy spark which makes the condition of the plug and its gap less critical.

SPARK PLUG

The spark plug is the part of the ignition system that ignites the fuel mixture, figure 7-28. It operates under severe and varying temperature conditions and is a critical part in engine operation.

All spark plugs are basically the same but they do differ in thread size, reach, heat range, and spark gap. There are hundreds of different types of spark plugs, each designed for the special requirements of a certain engine.

The *shell* of the spark plug is threaded so that it can easily be installed or removed from the cylinder head. Various spark plugs have different thread sizes. Some of the more common standard thread sizes are 7/8 inch, 10 millimeters, 14 millimeters, and 18 millimeters.

The *reach* of the spark plug is the distance between the gasket seat and the bottom of the spark plug shell, or about the length of the cut threads. The reach ranges from about 1/4 inch to 3/4 inch. Each engine must be equipped with a spark plug of the correct reach since reach determines how far the electrodes protrude into the combustion chamber. If the reach is too small, the electrodes have difficulty igniting the fuel when the spark jumps. If the reach is too great, the top of the piston may strike the electrodes on its upward stroke.

The *heat range* of a spark plug is the range of temperature within which the spark plug is designed to operate. If a spark plug operates

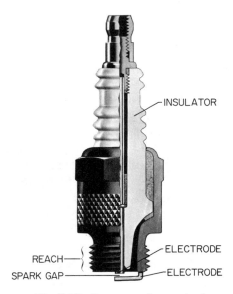

INSULATOR

REACH

ELECTRODE

SPARK GAP

ELECTRODE

Fig. 7-28 Cutaway of a spark plug

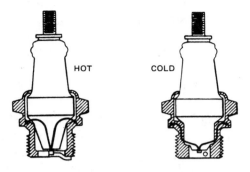

Fig. 7-29 Hot and cold spark plugs

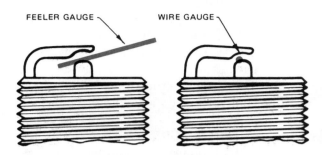

Fig. 7-30 A plain flat feeler gauge cannot accurately measure the true width of a spark gap.

at too low a temperature, it is quickly fouled with oil and carbon. If the spark plug operates at too high a temperature, the electrodes quickly burn up. The heat range for spark plugs depends on how fast the heat can be carried away from the electrodes. The path of heat transfer is from the electrodes to the ceramic insulator through the shell and into the cylinder head. The length of the ceramic insulator exposed to the combustion chamber determines the heat range of a spark plug. The longer this insulator, the longer the heat path and the hotter the spark plug's operating temperature. Likewise, the shorter this insulator, the shorter the heat path and the colder the spark plug's operating temperature, figure 7-29.

The *spark gap* (distance between the electrodes) must be correctly set. It is within this small space that the spark jumps and combustion begins. The gap must be large enough for sufficient fuel mixture to get between the electrodes but not so large as to prevent the spark from jumping across. The spark gap for various plugs ranges from about 0.020 inch to 0.040 inch.

If the spark plug is removed for cleaning, the gap should be checked and reset according to the manufacturer's specifications. A wire gauge should be used, figure 7-30.

BATTERY IGNITION

Battery ignition systems are often used on the larger types of small gasoline engines (4 to 15 horsepower), figure 7-31. With the battery a starter motor, or motor generator, can be used to crank the engine. This eliminates the need for various types of hand-pulled or wind-up starters. The battery can also be used for lights or other accessories, figure 7-32.

A typical battery ignition system includes virtually all of the magneto system parts that have been discussed earlier: the high-tension coil, breaker points, condenser, breaker cam, and spark plug. The one part that is replaced is the permanent magnet. The permanent magnet which supplies the energy to excite the coil is replaced by the battery. The most common systems for a small engine include a motor generator, battery, starter switch, voltage regulator, and the ignition system parts themselves, figure 7-33.

The starter generator or motor generator uses electrical energy from the battery as a motor to crank the engine for starting, figures 7-34 and 7-35. After starting, the unit acts as a generator to keep the battery charged. Electrical energy is changed to mechanical energy for starting and mechanical energy is changed to electrical energy for charging the

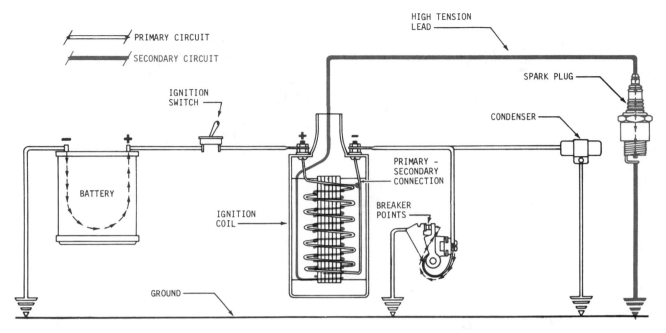

Fig. 7-31 Battery ignition system, primary (low voltage) and secondary (high voltage) circuits

Fig. 7-32 Applications that have lighting requirements and electric starters
are often equipped with a battery ignition system.

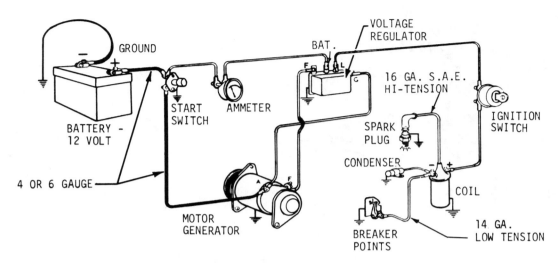

Fig. 7-33 Battery ignition system

battery. With the starter switch closed, the motor turns the crankshaft through a V-belt drive. Upon starting, the starter switch is released but the motor does not disengage. The V-belt drive continues but now the motor acts as a generator for recharging the storage battery. It should be noted that a typical auto system has a separate starter which disengages upon starting and a separate generator for charging the battery.

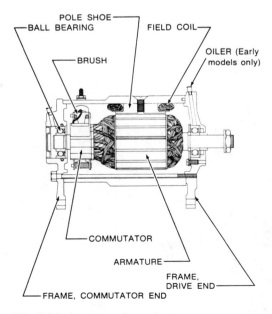

Fig. 7-34 Cutaway view of a motor generator

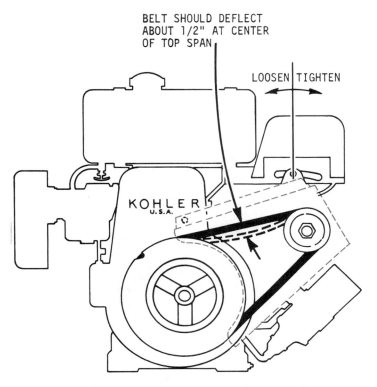

BELT SHOULD DEFLECT
ABOUT 1/2" AT CENTER
OF TOP SPAN

LOOSEN | TIGHTEN

KOHLER
U.S.A.

Fig. 7-35 Motor generator mounted on an engine

The storage battery may be either the 6-volt or 12-volt type, figure 7-36. The battery case contains alternate plates of lead peroxide and spongy lead, both active but chemically quite different. Separators prevent the plates from touching each other, figure 7-37. The battery is filled with a solution of sulfuric acid and water. Electrically speaking, the spongy lead plates are cathodes or the negative section, lead peroxide plates are the anodes or the positive section, and the sulfuric acid and water are the electrolyte. There is a voltage difference between the two groups of plates. When a complete circuit is made, current discharges from the battery. The circuit is closed and the electrolyte attacks the lead plates. The current is in motion and the battery is discharging. The

lead peroxide plates start to become coated with lead sulfide and begin to look like the negative plates. The sulfuric acid breaks down forming the lead sulfide and water. The water further dilutes the electrolyte. When the battery is discharged, the negative and positive plates are very similar in chemical structure. Voltage depends on the chemical difference between the two active materials. Figure 7-38 illustrates the electrochemical action in a battery.

Since the solution of sulphuric acid and water gets weaker or less dense as the battery discharges, the condition of the battery can be determined by checking the specific gravity of the solution with a battery hydrometer. The solution in a fully charged battery will have a specific gravity of 1.300. In a

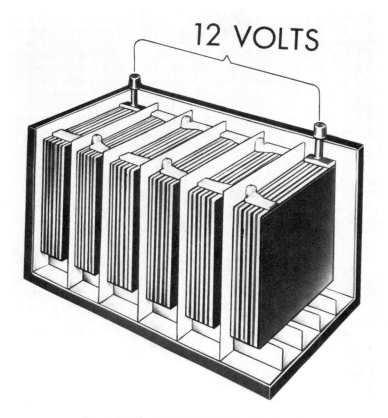

Fig. 7-36 Twelve-volt battery with six cells

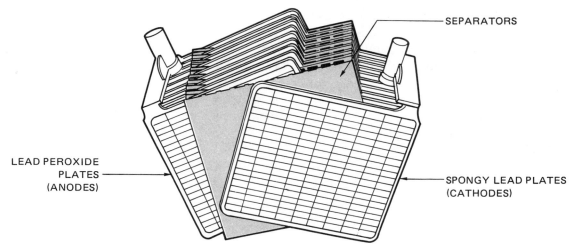

Fig. 7-37 Separators prevent the lead peroxide plates and spongy plates from touching each other

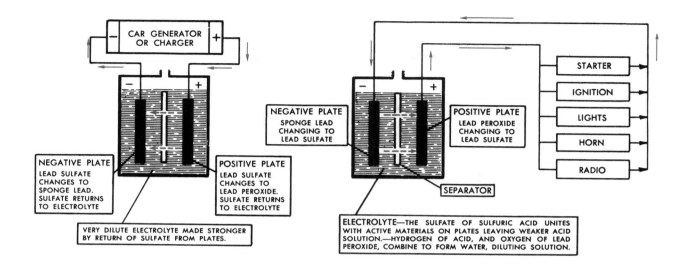

Fig. 7-38 Electrochemical action in a battery

discharged battery, the solution has a specific gravity of 1.120. Remember, the specific gravity of water is 1.000.

Fortunately, the action of the battery can be reversed. The starter generator recharges the battery without damage to the battery. The battery is recharged when the voltage produced by the generator is greater than battery voltage and the current being drawn from the battery is not greater than the input from the generator. Thus, the chemical action is reversed, and the battery is returned to its charged condition.

A current-voltage regulator comes into action when the engine has started and the starter switch is off and the engine is operating at 800 to 1000 rpm. The regulator prevents over charging the battery and also keeps the generator from producing voltages and current flow that could be damaging to the system.

REVIEW QUESTIONS

1. What three particles make up the atom?
2. Explain how electrons can move in a conductor.
3. What are the three ways that electricity can be produced or seen?
4. Explain the following electrical measurements: amperes, volts, and ohms.
5. State Ohm's Law.

6. Explain how permanent and temporary magnets differ.
7. What happens when a wire is cut by magnetic lines of force?
8. What happens to a coil of wire when electric current flows through it?
9. Explain how a transformer works.
10. In a magneto, what is the source of the magnetic field?
11. What are the two parts of the high-tension coil?
12. What purpose does the laminated iron core serve?
13. What is the function of the breaker points? How are they opened?
14. What is the function of the condenser? The spark plug?
15. What are the two electrical circuits in the magneto?
16. Why must the spark be advanced at high speeds?
17. What is the advantage of impulse coupling?
18. Can magneto ignition be used for multicylinder engines?
19. What parts of a conventional magneto are replaced on a solid state ignition system?
20. Explain the function of the transistor diodes in a solid state ignition system.
21. Explain how a silicon controlled rectifier acts as a switch or gate.
22. In what four ways do spark plugs differ?
23. What is the main advantage of having a battery in the ignition system?
24. a. When does the motor generator work as a motor?
    b. When does it work as a generator?
25. Storage battery plates are made of what material?
26. What is an electrolyte?

CLASS DISCUSSION TOPICS

- Discuss the structure of an atom.
- Discuss how conductors and insulators differ.
- Discuss the molecular theory of magnetism.
- Discuss the advantages of magneto ignition.
- Discuss the complete magneto cycle.

CLASS DEMONSTRATION TOPICS

- Demonstrate a magnetic field around a magnet using a permanent magnet, iron filings and paper.

- Demonstrate current flow caused by varying magnetic fields using a galvanometer, coil of wire, and permanent magnet.

- Demonstrate a magnetic field around a coil of wire using a direct-current source, coil of wire, iron filings, and paper.

- Disassemble a magneto showing the basic parts.

- Demonstrate how the breaker points are opened and show their separation at maximum opening (feeler gauge).

LABORATORY EXPERIENCE 7-1
BASIC FLYWHEEL MAGNETO PARTS

The magneto ignition system is a very reliable and trouble-free system. One big advantage of the magneto is that it does not require a storage battery for a power source. The magneto provides the spark for ignition: voltages of a few hundred volts are stepped up to about 20,000 volts which is enough to jump the spark gap in the spark plug.

There are seven basic parts which work together to produce the ignition spark:

1. high-tension coil

2. laminated iron core

3. condenser

4. breaker points

5. permanent magnets (on flywheel)

6. spark plug lead

7. spark plug

A brief review of the complete magneto cycle might be helpful in the study of the parts. When the permanent magnet is far away from the high-tension coil it has no effect on the coil. As the permanent magnet comes closer and closer, the primary coil feels the increasing magnetic field. The coil is being cut by magnetic lines of force and, therefore, there is current in the primary coil. This current travels through the breaker points and into ground. When the permanent magnet is just about opposite the high-tension coil the magnetic field around both coils is reaching its peak. Also, the piston is reaching the top of its stroke, compressing the fuel mixture.

The position of the permanent magnet is now causing the polarity of the laminated iron core and the coils to reverse direction. The reversing of direction is momentarily choked or held back by the coil. The breaker points then open interrupting the current flow and allowing the reversal and subsequent rapid collapse of the intense magnetic field that had been built up around the primary and secondary coils. Magnetic lines of force are cutting coils very rapidly and high voltages are induced in the coils. In the secondary coil the voltage may reach 18,000 to 20,000 volts which is enough to jump across the spark gap in the spark plug. This spark ignites the fuel mixture and the piston is forced down the cylinder.

STUDENT ASSIGNMENT

You are to disassemble the engine to expose flywheel magneto parts. Study the arrangement of the parts. Turn the engine over and observe how the breaker points are opened and closed. Remove the basic parts of the ignition system for closer inspection. It is essential that you record your work in a work record box as you complete each step of the disassembly procedure.

Disassembly Procedure

Instructors may want to supplement or revise specific steps of this procedure since there are many makes of engines. The following procedure is a general guide.

1. Remove the spark plug lead from the spark plug.

2. Remove any air shrouding, grass screens, or recoil starters from the flywheel area.

3. Remove the flywheel and flywheel nut. Use a flywheel puller if one is available. If a flywheel puller is not at hand, deliver a sharp blow to the end of the crankshaft with a plastic or soft hammer.

WORK RECORD BOX

PART	DISASSEMBLY (nuts, bolts, etc.)	OPERATION PERFORMED	TOOL USED

4. Remove the breaker point cover (on some engines).

5. Turn the crankshaft over and observe the opening and closing of the breaker points.

6. Remove the high-tension coil and laminated iron core.

7. Remove the condenser.

8. Remove the breaker point assembly.

9. Remove the breaker cam (on some engines).

Reassembly Procedure

Reverse the disassembly procedure. If the engine is to be operated upon reassembly, several points are very important.

1. Reinstall the breaker cam in the same way as it came off.

2. Use care with the key and the keyways; do not force.

3. Breaker points must be reset to manufacturer's specifications.

4. The laminated iron core must be positioned for the proper air gap between it and the flywheel magneto.

5. All connections should be checked for tightness.

REVIEW QUESTIONS

Study the flywheel magneto system and then answer the following questions.

1. List the magneto parts in the primary circuit.

2. List the magneto parts in the secondary circuit.

3. Explain how the spark is triggered.

4. On which engine stroke is the spark provided?

5. What opens and closes the breaker points?

6. What happens to the primary electrical circuit when the breaker points open?

7. What happens to the magnetic field when the breaker points open?

8. What is the purpose of using a laminated iron core instead of a solid core?

9. What are the functions of the condenser?

10. Discuss briefly what happens in the magneto cycle with one revolution of the crankshaft.

LABORATORY EXPERIENCE 7-2
SOLID STATE IGNITION

The solid state ignition system uses electronic components such as transistor diodes and silicon controlled rectifiers instead of breaker points, breaker cam, and the condenser. The flywheel magnet is the only moving part of this ignition system.

The electronic devices are sealed in a module mounted near the flywheel. The module is very reliable and since it is a sealed unit it cannot be taken apart.

STUDENT ASSIGNMENT

You are to disassemble the engine to expose the solid state ignition parts. Study the arrangement of the parts and locate the trigger module, iron core, and coils.

Disassembly Procedure

1. Remove the spark plug lead from the spark plug.

2. Remove any air shrouding, grass screens, recoil starters, etcetera from the flywheel area.

WORK RECORD BOX

PART	DISASSEMBLY (nuts, bolts, etc.)	OPERATION PERFORMED	TOOL USED

3. Locate the solid state ignition parts.

NOTE: If solid state ignition parts are removed from the engine, they must be replaced exactly as they came off. Check the mechanic's repair book for exact specifications.

Reassembly Procedure

Reverse the disassembly procedure. The air gap between the flywheel and the iron core is critical.

REVIEW QUESTIONS

1. Prepare a sketch of the module and other ignition system parts on your engine.

UNIT 8

Routine Care and Maintenance, and Winter Storage

OBJECTIVES After completing this unit, the student should be able to:
- Discuss the important areas of routine care and maintenance.
- Identify types of spark plug fouling.
- Clean and regap spark plugs.
- Discuss the various types of air cleaners and their care.
- Correctly prepare an engine for winter storage.
- Locate routine care and maintenance information in an owner's instruction book.

A gasoline engine represents an investment. To safeguard this investment the engine operator must perform certain routine steps in care and maintenance. Routine care and maintenance help to ensure the longest life possible for engine parts and can save on repair bills.

The best source of information on engine care is the owner's manual or operator's instruction book that comes with every new engine. The book contains the information that the manufacturer considers necessary for the owner to know. The information included allows the owner to obtain top performance and long engine life. Some typical topics discussed are lubrication procedures, carbure-

tor adjustment, ignition system data, air cleaner cleaning instructions, winter storage, and care of the machine that the engine is used on. The information in the book is important. A person with a new engine should read and thoroughly understand the instruction book before touching the engine. A plastic cover helps to keep the book in good condition.

ROUTINE CARE AND MAINTENANCE

Routine care and maintenance is given to an engine during its normal use. It is provided to keep the engine operating at its peak

Service	Daily (Pre-start)	Every 25 Hours	Every 50 Hours	Every 100 Hours	Every 500 Hours
Clean Air Intake Screen	X				
Check Oil Level.	X				
Replenish Fuel Supply	X				
Change Lube Oil		X			
Check Air Cleaner (Replace if Dirty).			X		
Clean Cooling Fins and External Surfaces.			X		
Service or Replace Spark Plugs.				X	
Have Breaker Points Checked and Serviced*					X
Have Ignition Timing Checked*					X
Have Valve — Tappet Clearance Checked*					X
Have Service Cylinder Heads Serviced*					X†

*Have these services done only by qualified engine specialists.

†If leaded gasoline is used service cylinder head every 250 hrs.

NOTE: Intervals stated are for good, clean operating conditions only — service more frequently (even daily) if extremely dusty or dirty conditions prevail.

Fig. 8-1 All engines should have a service schedule similar to this one for a Kohler Model K361.

efficiency and to prevent undue wear of engine parts. There are four basic points to consider: oil supply, cooling system, spark plug, and air cleaner.

Oil Supply

On a four-cycle engine, the oil supply in the crankcase should be checked daily or each time the engine is used. If the oil level has dropped below the *add* mark on the dipstick, fill the crankcase to the proper level. If the engine does not have a dipstick, fill the crankcase until the oil can be seen in the filler hole. A new engine or one that is being used for the first time must be carefully checked to be sure that oil is in the crankcase.

Most manufacturers recommend that the oil be changed after a short period of time when breaking in a new engine. Depending on the manufacturer, this first oil change should be after two to twenty hours of operation.

After the oil has been changed once, the engine can go for longer periods of time be-tween oil changes. Lauson specifies changing engine oil after every 10 hours of operation; Briggs and Stratton after every 20 hours; Kohler after every 25 hours; Gravely and Wisconsin after every 50 hours, Onan after every 100 hours. Each manufacturer's instruction book gives the length of time between oil changes and the exact SAE number and quality of oil to be used.

Two-cycle engines have their lubricating oil mixed with the gasoline. The owner's manual specifies the proportions of oil to gasoline and the type of oil to be used.

Cooling System

The most important consideration regarding the cooling system is to keep it clean. On air-cooled engines, clean the radiating fins, flywheel vanes, and shrouds whenever they begin to accumulate grass, dirt, etc., figure 8-2. Do not allow deposits to build up which reduce the engine's cooling capacity. How often the cooling system is cleaned depends

Use length of stiff wire to dislodge accumulated dirt around cylinder fins.

Avoid damage to fins.

Fig. 8-2 Keep heat radiating fins clean

on the dust conditions under which the engine operates.

On a water-cooled engine, check the water level in the radiator before operating the engine. If the water level is low, fill it to the correct level.

Outboard motors need little routine care for the cooling system. However, each time the engine is started the operator should check to see that the engine is pumping water. This can be detected on most engines by a small amount of water coming from the telltale holes. If the engine is not pumping water, it should be stopped immediately and the source of the trouble corrected. Do not put hands by the underwater discharge to determine if the engine is pumping water as the propeller may be dangerously close. If the engine's cooling system is thermostatically controlled, it may take a few moments for the thermostat to open to allow a flow of water from the engine.

Spark Plugs

Spark plugs need to be cleaned and regapped periodically. Spark plugs are usually cleaned after 100 hours of operation; some manufacturers recommend cleaning after as little as 50

hours. Since spark plugs operate under severe conditions, they become fouled easily. Knowing the signs of fouling can often tell about the engine's overall condition. Common problems that develop with spark plugs are carbon fouling, burned electrodes, chipped insulator, and splash fouling, figure 8-3.

Normal spark plugs have light tan or gray deposits but show no more than 0.005 inch increase in the original spark gap. These can be cleaned and reinstalled in the engine.

Worn out plugs have tan or gray colored deposits and show electrode wear of 0.008 inch to 0.010 inch greater than the original gap. Throw away such plugs and install new ones.

Oil fouling is indicated by wet, oily deposits. This condition is caused when oil is pumped (by the piston rings) to the combustion chamber. A hotter spark plug can help but the condition may have to be corrected by engine overhaul.

Gas fouling or fuel fouling is indicated by a sooty, black deposit on the insulator tips, electrodes, and shell surfaces. The cause may be a too rich fuel mixture, light loads, or long periods at idle speed.

Carbon fouling is indicated if the plug has dry, fluffy, black deposits. This condition may be caused by a too rich fuel mixture, improper carburetor adjustment, choke partly closed, or clogged air cleaner. Also, slow speeds, light loads, long periods of idling, and the resulting cool operating temperature can cause deposits not to be burned away. A hotter spark plug may correct carbon fouling.

Lead fouling is indicated by a soft, tan, powdery deposit on the plug. These deposits of lead salts build up during low speeds and light loads. They cause no problem at low speed but at high speeds, when the plug heats up, the fouling often causes the plug to misfire, thus limiting the engine's top performance.

NORMAL SPARK PLUG

WORN OUT SPARK PLUG

CARBON--FOULED SPARK PLUG

CHIPPED INSULATOR ON SPARK PLUG

OVERHEATED SPARK PLUG

OIL--FOULED SPARK PLUG

SPLASH-FOULED SPARK PLUG

BENT SIDE ELECTRODE

Fig. 8-3 Normal and damaged spark plugs

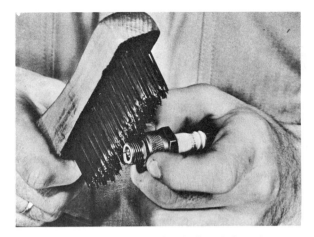

Fig. 8-4 Wire brush the shell and threads

Burned electrodes are indicated by thin, worn away electrodes. This condition is caused by the spark plug overheating. This overheating can be caused by lean fuel mixture, low octane fuel, cooling system failure, or long periods of high-speed heavy load. A colder plug may correct this trouble.

Splash fouling can occur if accumulated cylinder deposits are thrown against the spark plugs. This material can be cleaned from the plug and the plug reinstalled. If spark gap tools or pliers are not properly used, the electrode can be misshaped into a curve.

Cleaning Spark Plugs.   If inspection indicates that the spark plug needs cleaning, it can be cleaned within a few minutes. First, wire brush the shell and threads not the insulator and electrodes, figure 8-4. Then wipe the plug with a rag that has been saturated with solvent. This removes any oil film on the plug. Next, file the sparking surfaces of the electrodes with a point file, figure 8-5. Finally, regap the electrodes setting them according to specifications, figure 8-6. Check the gap with a wire spark gap gauge, figure 8-7. Abrasive blast cleaning machines found

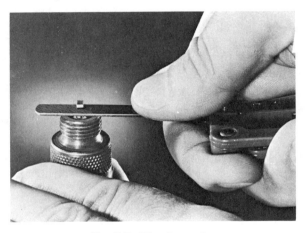

Fig. 8-5 File electrodes

Fig. 8-6 Regap the plug

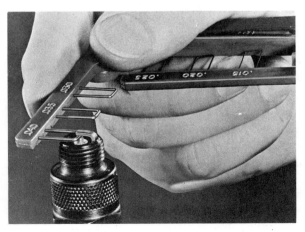

Fig. 8-7 Check the spark gap

Fig. 8-8 About this much abrasive material would enter a six-cylinder engine every hour if the air cleaner was not used.

in many service stations are not recommended by several engine manufacturers. Do not allow any foreign material to fall into the cylinder while the spark plug is out. Also, do not forget the spark plug gasket when reinstalling the spark plug. This copper ring provides a perfect seal.

Cleaning cannot repair a broken, cracked, or otherwise severely damaged spark plug. Cleaning is designed to maintain good operation and prolong spark plug life. Many small engine manufacturers recommend the installation of a new spark plug at the beginning of

each season. On a single-cylinder engine, the spark plug is more critical than on a six- or eight-cylinder engine.

Air Cleaner

The air cleaner serves the important function of cleaning the air before it is drawn through the carburetor and into the engine. Small abrasive particles of dust and dirt are trapped in the air cleaner, figure 8-8. The air cleaner should be periodically cleaned; how often depends mainly on the atmosphere in which the engine is operating, figure 8-9. In a dusty atmosphere, (a garden tractor used in a dry garden for example), the air cleaner should be cleaned every few hours. Under normal conditions, the air cleaner should be cleaned every 25 hours or sooner.

Air cleaners can be classified as either oil type or dry type. Oil type includes oil-bath air cleaners and oil-wetted polyurethane (foam) filter element air cleaners. Dry type includes foil, moss, or hair element, felt or fiber hollow element, and metal cartridge air cleaners.

Fig. 8-9 Engines operating in different environments have different air cleaner requirements.

Fig. 8-10 Oil-bath air cleaner

Oil-type Air Cleaners

Oil-bath Air Cleaner.   The oil-bath air cleaner carries a small amount of oil in its bowl, figure 8-10. The level of this oil should be checked before each use of the engine. If the oil level is low, fill it to the mark but not above the mark. Use the same type of oil as is used in the crankcase. As this cleaner works, both the oil and filter element become dirty. To clean an oil-bath air cleaner, disassemble the unit and then pour out the dirty oil. Wash the bowl, cover, and filter element in solvent. Dry all parts. Refill the bowl to the correct level and then reassemble the unit, figure 8-11.

Oil-wetted Polyurethane Filter Element Air Cleaner.   Shown in figure 8-12, this type of air cleaner is perhaps the most common. Clean it by washing the element in kerosene, liquid detergent and water, or an approved solvent. Dry the element by squeezing it in a cloth or towel. Apply considerable oil to the foam and work it throughout the element. Squeeze out any excess oil and then reassemble the air cleaner.

Fig. 8-11 Clean, fill, reassemble air cleaner

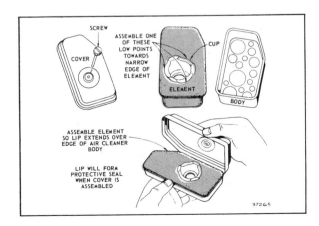

Fig. 8-12 Oil foam air cleaner

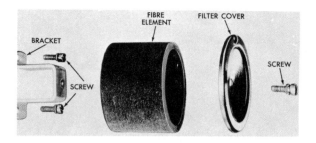

Fig. 8-13 Fiber-element type air cleaner

Dry-type Air Cleaners

Foil, Moss, or Hair Element Air Cleaner. This type of element is also washed in solvent. After washing, allow the element to dry, then return it to the air cleaner body.

Felt or Fiber Cylinder Element Air Cleaners. These air cleaners should be cleaned by blowing compressed air through them in the opposite direction from normal flow, figure 8-13. Dirt and dust particles will be blown out of the element.

Metal-cartridge Air Cleaners. These air cleaners are cleaned by tapping or shaking to dislodge dirt accumulations, figure 8-14.

Air cleaner elements are replaceable parts. If they wear out or become very clogged, they should be thrown away and replaced with a new filter element. As a simple test to determine if the air cleaner is badly clogged: (1) run the engine with the air cleaner removed; (2) then, with the engine still running, replace the air cleaner; and (3) notice if the engine speed remains constant or drops down. Any noticeable drop in speed probably indicates that the filter element is clogged.

Periodically inspect the engine for loose parts. Do not neglect to lubricate the engine's associated linkages, gear reduction units, chains, shafts, wheels, or the machinery the engine powers. These requirements vary greatly from engine to engine. On an outboard engine remember that the lower unit requires special lubrication.

EXHAUST PORTS (TWO-CYCLE)

Cleaning the exhaust ports of a two-cycle engine can also be considered a part of routine care and maintenance, figure 8-15. Over a period of time these ports can become clogged with carbon deposits which greatly reduce power and performance. After removing the muffler to expose the ports, the

Fig. 8-14 Metal-cartridge type air cleaner

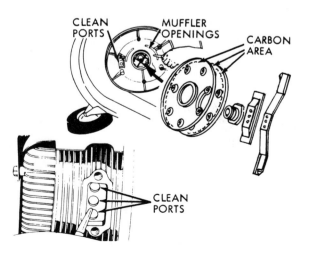

Fig. 8-15 Cleaning the exhaust ports

carbon can be cleaned out with a soft object such as a dowel rod. Be sure to move the piston so that it covers the ports. This prevents loose deposits from going into the cylinder.

WINTER STORAGE AND CARE

Preparing an engine for winter storage at the end of the season should be done carefully. Several procedures should be done:

- Drain the fuel system.

- Inject oil into the cylinder (upper cylinder).

- Drain the crankcase (four-cycle).

- Clean the engine.

- Wrap the engine in a canvas or blanket and store in a dry place.

The entire fuel system should be drained: tank, fuel lines, sediment bowl, and carburetor. Gasoline that is stored over a long period of time has a tendency to form gummy deposits or varnish. These deposits could injure and block the fuel system if gasoline stands in the system for a long time. Do not plan to hold gasoline over from season to season. Purchase a new supply when the engine is returned to use.

Many manufacturers recommend that a small amount of oil be poured into the cylinder prior to storage. Remove the spark plug and pour in the oil (about a tablespoon in most cases). Turn the engine over by hand a few times to spread the oil evenly over the cylinder walls. Replace the spark plug.

Some manufacturers suggest that the oil be introduced into the cylinder while the engine is running at a slow speed and just before the engine is stopped. To do this, the air cleaner is removed and a small amount of oil is poured in. The oil goes through the carburetor and then into the combustion chamber. When dark blue exhaust is produced the engine can be stopped.

The crankcase should be drained while the engine is warm so the oil drains more readily. In some cases the crankcase is then refilled; on most engines it need not be refilled until the engine is returned to service.

Any unpainted surfaces that might rust should be lightly coated with oil before storage. Linkages should be oiled. The radiating fins and the engine in general should be cleaned. Do not put too much oil on the engine as it will collect dirt and be difficult to remove.

Finally, wrap the engine in a canvas or dry blanket and store the engine in a dry place. A garage, dry shed, or dry basement is good. Do not allow the engine to be left outside, exposed to the weather.

RETURNING THE ENGINE TO SERVICE

Upon returning the engine to service refill the gas tank with new gasoline. Before filling the gas tank check to see if any water has condensed in the tank. If so, drain the water completely.

Check for condensation in the crankcase and refill the crankcase with new oil of the correct type, figure 8-16. Also, clean the air cleaner and refill the bowl with oil if it is an oil-bath type.

Check the spark plug; clean and regap it. Many manufacturers of small engines recommend that a new plug be installed at the beginning of each season to ensure peak performance throughout the season.

(A) Fill the gas tank

(D) Choke closed

(B) Check the oil in the crankcase

(E) Spark plug shorting bar off spark plug

(C) Fuel shut-off valve open

(F) Air cleaner clean

Fig. 8-16 Things to check before returning the engine to service

SPECIAL CONSIDERATIONS FOR OUTBOARD MOTORS

Winter Storage

Generally, winter storage for outboard motors is similar to that for other engines. One notable difference is the cooling system. Being water cooled, all water must be removed from the cooling system. If a water-filled cooling system on an idle engine is exposed to freezing weather, cracked water jackets and other major damage can result.

To remove the water from the system, remove the engine from the water, set the speed control on stop (to prevent accidental starting), and turn the engine over several times by hand. This allows the water to drain from the water jackets and passages.

Salt Water Operation

Although most engines are treated with anticorrosives, not all engines are free from corrosion. If an engine normally used in salt water is not to be used for a while, its cooling system should be flushed with fresh water to prevent any possible corrosion.

Use in Freezing Temperatures

If the engine is being used during freezing temperatures, care must be taken not to allow the cooling water to freeze in the engine or lower unit during an idle period. While the engine is operating there is no danger of freezing.

REVIEW QUESTIONS

1. Explain what routine care and maintenance is and why it is important.
2. Where can routine care and maintenance information be found?
3. What are the four basic points to be considered in routine care and maintenance?
4. How often should the oil in the crankcase be checked?
5. How often should the oil in the crankcase of a Briggs and Stratton engine be changed?
6. What are the two types of cooling systems?
7. How often should the air-cooling system be cleaned?
8. How is an air-cooling system cleaned?
9. What care should be given a water-cooling system?
10. How often should spark plugs be cleaned?
11. What is the purpose of cleaning a spark plug?
12. List several types of spark plug fouling.
13. What are the steps in cleaning a spark plug?
14. What is the purpose of the air cleaner?

15. How often should air cleaners be cleaned?

16. What are the two basic types of air cleaners?

17. Discuss the cleaning of the several types of air cleaners.

18. List the general steps in winter storage of engines.

19. List the general steps in returning the engine to service.

20. What special considerations must be made for outboard motors?

CLASS DISCUSSION TOPICS

- Discuss the importance of routine care and maintenance.
- Discuss the importance of care in winter storage.
- Discuss the types of cooling systems and their maintenance.
- Discuss the types of spark plug fouling.
- Discuss the importance and function of the air cleaner.
- Discuss the various types of air cleaners and how each might be used.
- Discuss the importance of safety with gasoline and gasoline engines.
- Discuss state laws in regard to gasoline storage.

CLASS DEMONSTRATION TOPICS

- Demonstrate cleaning and care of cooling systems.
- Demonstrate cleaning spark plugs.
- Demonstrate cleaning air cleaners.
- Demonstrate putting an engine into winter storage.
- Demonstrate returning an engine to service after winter storage.
- Demonstrate the check-out of an engine prior to starting.
- Demonstrate routine care and maintenance on an outboard engine.
- Demonstrate routine care and maintenance on a four-stroke cycle engine.

LABORATORY EXPERIENCE 8-1
CLEAN AND REGAP SPARK PLUGS

Spark plug performance can often be improved by periodic cleaning. Also spark plug life can be lengthened. Most manufacturers of engines recommend that spark plugs be cleaned every 100 hours of engine operation. If a plug is damaged, insulator cracked, electrodes worn away, etc., cleaning cannot repair the damage.

Since spark plugs operate under severe conditions they become fouled easily. Knowing the signs of fouling can often tell about the engine's overall condition. Common problems that develop with spark plugs are: carbon fouling, oil fouling, gas fouling, lead fouling, burned electrodes, chipped insulator, and splash fouling.

Normal spark plugs have light tan or gray deposits but show no more than 0.005" increase in the original spark gap. These can be cleaned and reinstalled in the engine.

Worn out plugs have tan or gray colored deposits and show electrode wear of 0.008" to 0.010" greater than the original gap. Throw this plug away and install a new plug.

Oil fouling is indicated by wet, oily deposits. This condition is caused when oil is pumped (by the piston rings) to the combustion chamber. A hotter spark plug may help but the condition may have to be remedied by engine overhaul.

Gas fouling or fuel fouling is indicated by a sooty, black deposit on the insulator tips, the electrodes and shell surfaces. The cause may be excessively rich fuel mixture, light loads, or long periods at idle speed.

Carbon fouling is indicated if the plug has dry, fluffy, black deposits. This condition may be caused by excessively rich fuel mixture; improper carburetor adjustment, choke partly closed, clogged air cleaner. Also slow speeds, light loads, long periods of idling and the resulting cool operating temperature may cause deposits not to be burned away. A hotter spark plug may correct carbon fouling.

Lead fouling is indicated by a soft, tan, powdery deposit on the plug. These deposits of lead salts build up during low speeds and light loads. They cause no problem at low speed but at high speeds, when the plug heats up, the fouling often causes the plug to misfire, thus limiting the engine's top performance.

Burned electrodes are indicated by thinned out, worn away electrodes. This condition is caused by the spark plug overheating. This overheating can be caused by lean fuel mixture, low octane fuel, cooling system failure, or long periods of high-speed heavy load. A colder plug may correct this trouble.

Splash fouling may occur if accumulated cylinder deposits are thrown against the spark plugs. This material can be cleaned from the plug and the plug reinstalled.

If spark gap tools and/or pliers are not properly used, the electrode can be misshaped into a curve. Correct spark plug type and correct spark plug gap setting are found in the engine operator's manual.

STUDENT ASSIGNMENT

You are to remove the engine spark plug, examine it for damage and fouling, clean the plug and regap it. It is essential that you record your work in a work record box as you complete each step of the procedure.

Procedure

1. Remove the spark plug and examine the condition of the plug.
2. Check the spark gap before cleaning and resetting.
3. Wire brush the shell and threads.
4. Clean the insulator with a rag soaked in solvent.
5. Clean electrodes and sparking surfaces.
6. Regap the plug to the correct setting.
7. Reinstall the plug. (Do not forget the spark plug gasket.)

WORK RECORD BOX

PART	DISASSEMBLY (nuts, bolts, etc.)	OPERATION PERFORMED	TOOL USED

REVIEW QUESTIONS

Upon completion of the work, answer the following questions.

1. What is the spark plug type (commercial designation)?
2. What was the spark gap prior to cleaning?
3. Does the manufacturer suggest any equivalent spark plug types? If so, list them.

4. What is the correct spark gap for the plug?
5. Describe the condition of the plug prior to cleaning.
6. Explain the purpose of cleaning spark plugs.

LABORATORY EXPERIENCE 8-2
WINTER STORAGE OF ENGINES

Winter storage is an important but often neglected aspect of engine care. A few minutes of time spent at preparing the engine may save hours of engine trouble when the engine is returned to service. Fuel system, lubrication system, and cooling system need attention during winter storage.

STUDENT ASSIGNMENT

You are to prepare an engine for winter storage (either two-cycle or four-cycle). It is essential that you record your work in a work record box as you complete each step of the procedure.

Procedure

Your instructor may supplement or revise specific steps of the procedure which follows since there are many makes of engines. The following procedure is a general guide:

1. Remove spark plug lead.
2. Drain the fuel system: tank, fuel lines, sediment bowl, and carburetor. Clean the air cleaner.

WORK RECORD BOX

PART	DISASSEMBLY (nuts, bolts, etc.)	OPERATION PERFORMED	TOOL USED

3. Inject oil into the cylinder (upper cylinder). Remove the spark plug and pour in about 1 ounce of oil. Turn the engine over several times to spread the oil.

4. Drain the crankcase (four-cycle).

5. Clean the engine. Pay special attention to the cooling system.

6. Wrap the engine in a canvas or blanket and store in a dry place.

REVIEW QUESTIONS

Answer the following questions.

1. Explain why winter storage is important.

2. Should gasoline be kept over for the next season? Explain.

3. Is winter storage a job that can be done by an engine owner?

4. Why should the air cleaner be cleaned before storage instead of when the engine is returned to service?

5. List several good places to store an engine.

LABORATORY EXPERIENCE 8-3
INSTRUCTION BOOK USAGE

The manufacturer's instruction book is included with every new engine. The book contains the information that the manufacturer considers necessary for the owner to know. The information included enables the owner to obtain top performance and long engine life. Some typical topics that are discussed are: lubrication procedures, carburetor adjustment, ignition system data, air cleaner cleaning instructions, winter storage, and care of the machine that the engine is used on. The information in the book is important; a person with a new engine should read and thoroughly understand the instruction book before touching the engine. All too often the inexperienced person adopts the poor attitude, "if all else fails, read the instruction book." The book is valuable, read it, understand it, place it in a safe place for future reference.

Some manufacturers issue very small instruction books. Other manufacturers issue instruction books that are very inclusive, almost a repair manual. Regardless of the size of the book, it is an important piece of literature for the engine owner.

STUDENT ASSIGNMENT

You are to read an instruction book (one of your own or one the instructor gives you) and then fill in the requested information that follows. Don't be disturbed if you are not able to answer all of the questions. Every manufacturer has a slightly different idea on just how much information the engine owner should have to operate the engine.

NOTE: See figure 8-17 for a sample engine identification.

Engine Manufacturer's Name
Engine Model Number
Two- or Four-cycle Engine
Rated Horsepower at rpm

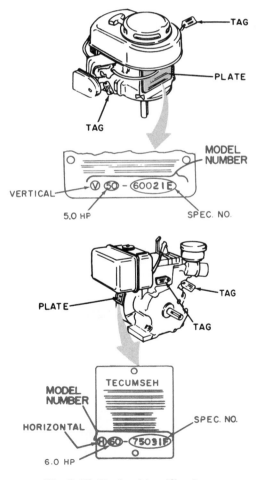

Fig. 8-17 Engine identification

Bore   Stroke   Displacement

Lubrication System:

Summer Oil Type

Winter Oil Type

Oil Checked — How Often

Oil Changed — How Often

Ignition System:

Spark Plug Type

Spark Plug Gap

Spark Plug Cleaned — How Often

Breaker Points Checked — How Often

Breaker Points Set To

Air Cleaner:

Dry Type or Oil Bath

Cleaned — How Often

Air Cleaner Cleaned With

Cooling System:

Air-Cooled or Water-Cooled

Cleaned — How Often (Air-Cooled)

Water Level Checked (Water-Cooled) — How Often

NOTE: Most questions are answered in terms of hours between cleaning, changing, setting, checking, and so forth.

LABORATORY EXPERIENCE 8-4
CLEANING THE AIR CLEANER

Care of the engine's air cleaner is important since all air that enters the engine goes through the air cleaner. A clogged air cleaner restricts air flow. A damaged air cleaner may allow dirt to enter the engine. The frequency of cleaning really depends on the atmosphere in which the engine is operated. Dusty conditions require frequency cleaning of the air cleaner. Clean atmospheric conditions enable the air cleaner to go for a longer time between cleaning.

STUDENT ASSIGNMENT

You are to disassemble, inspect, clean, and reassemble the air cleaner. After completing the work, record it in a work record box.

Cleaning Procedure

The engine instruction book is the final word on the correct cleaning of the air cleaner. If an instruction book is not available, the following procedure can be used as a general guide. Be careful not to allow any dirt to fall into the carburetor when the air cleaner is removed.

A. Oil-Bath Air Cleaner
    1. Loosen and carefully remove the air cleaner from the carburetor.
    2. Wash out the filter element in solvent.
    3. Pour the old oil out of the bowl and wipe the bowl clean.
    4. Refill the bowl to the correct level. (Use the same type oil as that in the crankcase.)
    5. Reassemble the air cleaner.

B. Oil-Wetted Polyurethane (Foam) Filter Element (See figure 8-18.)
    1. Loosen and carefully remove the air cleaner from the carburetor.
    2. Wash out the filter element in solvent.
    3. Squeeze out excess solvent.
    4. Work about one teaspoon of oil into the filter.
    5. Reassemble the air cleaner.

C. Foil, Moss, or Hair Elements; Dry Type
    1. Loosen and carefully remove the air cleaner from the carburetor.
    2. Wash out the filter element in solvent, and allow it to dry.
    3. Reassemble the air cleaner.

D. Felt or Fiber Cylinder Element; Dry Type
    1. Loosen and carefully remove the air cleaner from the carburetor.
    2. Blow out the element with compressed air; reverse direction of the normal flow.
    3. Reassemble the air cleaner.

E. Metal Cartridge Air Cleaner; Dry Type
    1. Loosen and carefully remove the air cleaner from the carburetor.
    2. Tap or shake the air cleaner to dislodge dirt and dust.
    3. Reassemble the air cleaner.

A Wash the element

C Pour oil on the element

B Dry the element

D Work oil into the element

Fig. 8-18 Cleaning polyurethane foam air cleaner

WORK RECORD BOX

PART	DISASSEMBLY (nuts, bolts, etc.)	OPERATION PERFORMED	TOOL USED

REVIEW QUESTIONS

Upon completion of the work, answer the following questions.

1. Explain what would happen to the fuel mixture if the air cleaner gradually became clogged.

2. What harmful effect can dirty air have on an engine?

3. Classify the following operations in relation to atmospheric conditions (dusty, average, or clean):
   a. Engine operated in feed mill
   b. Engine on small snow plow
   c. Engine mulching dry leaves
   d. Outboard motor on lake
   e. Engine on garden tractor

UNIT 9

Troubleshooting, Tune-up, and Reconditioning

OBJECTIVES   After completing this unit, the student should be able to:
- Correctly clean and adjust the breaker points, and test the spark produced at the spark plug.
- Test the magneto output, and correctly adjust engine timing.
- Correctly replace piston rings, and correctly lap valves.
- Check the compression of an engine.
- Correctly use the feeler gauge and torque wrench.
- Perform an engine tune-up.

Troubleshooting is the intelligent, step-by-step process of locating engine trouble. The troubleshooter examines and/or tests the engine to determine the cause of its disorder. To engage in this process, a person must thoroughly understand the how and why of engine operation. Troubleshooting is more a mental activity than a physical one.

The troubleshooter first establishes the engine's symptoms. The most common causes for these symptoms are then checked. If the solution is not found, another possible cause for the engine's problem is explored. The possible cause for engine failure must quickly be narrowed down because there are many afflictions that can creep into an ailing engine.

An experienced mechanic troubleshoots quickly and with little apparent effort. Years of experience and a thorough understanding of engine operation allow the mechanic to work with such competence.

Engine owners can do their own troubleshooting when they understand engine operating principles. Many owner's manuals contain a troubleshooting chart that is of great assistance. With the aid of an engine troubleshooting chart, the engine owner or operator can locate and correct many engine difficulties without calling in a professional mechanic.

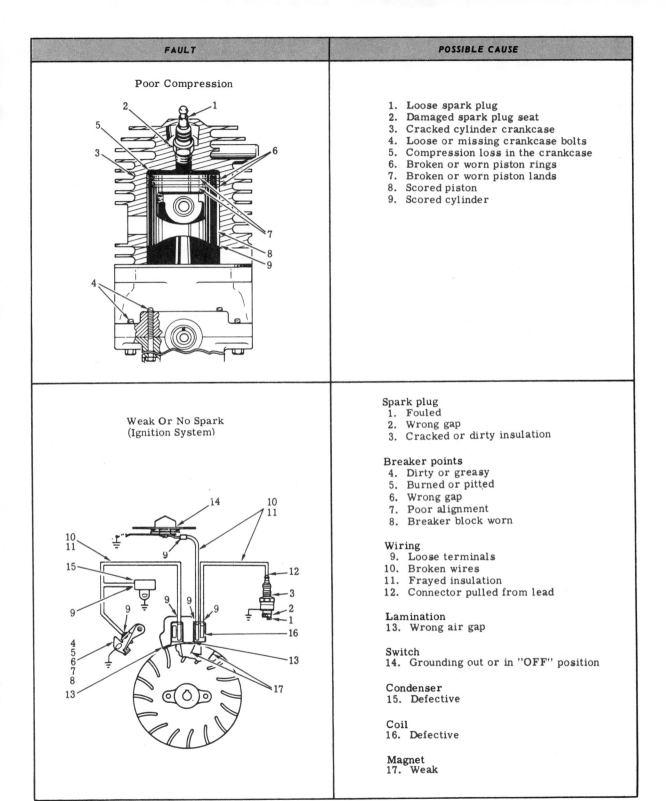

FAULT	POSSIBLE CAUSE
Poor Compression	1. Loose spark plug 2. Damaged spark plug seat 3. Cracked cylinder crankcase 4. Loose or missing crankcase bolts 5. Compression loss in the crankcase 6. Broken or worn piston rings 7. Broken or worn piston lands 8. Scored piston 9. Scored cylinder
Weak Or No Spark (Ignition System)	**Spark plug** 1. Fouled 2. Wrong gap 3. Cracked or dirty insulation **Breaker points** 4. Dirty or greasy 5. Burned or pitted 6. Wrong gap 7. Poor alignment 8. Breaker block worn **Wiring** 9. Loose terminals 10. Broken wires 11. Frayed insulation 12. Connector pulled from lead **Lamination** 13. Wrong air gap **Switch** 14. Grounding out or in "OFF" position **Condenser** 15. Defective **Coil** 16. Defective **Magnet** 17. Weak

Fig. 9-1 Troubleshooting chart

Troubleshooting requires a knowledge of engine operation as well as some common sense. For example, if the engine does not start, there are two possible reasons: either (1) fuel is not getting into the combustion chamber, or (2) fuel is being provided but it is not being ignited. Therefore, the trouble is either in the fuel system or the ignition system. Lubrication or cooling systems have little influence on the ability of the engine to start so these systems are not inspected.

The troubleshooter starts with the simplest and most frequent causes of trouble. In inspecting the fuel system, the first step is to check for fuel in the tank. If this is not the cause of the trouble, the fuel shutoff valve is then checked. The inspection continues in a logical manner until the trouble is found. It would be a mistake to completely disassemble the carburetor or magneto as the first step in troubleshooting.

Look at the troubleshooting charts (figures 9-1 to 9-3) and discover how many of the common engine troubles can be corrected by an owner who has the tools on hand.

ENGINE TUNE-UP

Engine tune-up does not involve major engine repair work. Rather, it is a process of cleaning and adjusting the engine so that it gives top performance. A tune-up can be done by an experienced engine owner or it can be done by a mechanic. Before beginning tune-up THINK SAFETY. Study the following safety rules.

GENERAL SAFETY RULES FOR TUNE-UP AND RECONDITIONING

- Do not run an engine in a closed building without proper ventilation. Exhaust fumes contain carbon mon—oxide which is poisonous and can be fatal.
- Do not touch the cylinder head, cylinder block, or exhaust system on an engine that is running or one that has just been stopped. The parts may be hot enough to cause a severe burn.

Problem	No Fuel	Improper Fuel	Dirt In Fuel Line	Dirty Air Screen	Incorrect Oil Level	Engine Over-Loaded	Dirty Filter Element	Faulty Spark Plug
Will not start	X		X			X	X	X
Hard starting	X	X	X	X		X		X
Stops suddenly	X		X	X	X	X	X	
Lacks power		X	X	X		X	X	X
Operates erratically		X	X	X		X	X	X
Knocks or pings		X		X		X		X
Skips or misfires		X	X	X			X	X
Backfires			X			X	X	X
Overheats			X	X	X	X	X	
High fuel con-sumption							X	X

Fig. 9-2 Troubleshooting chart for the engine owner

Engine Will Not Crank

Hydrostatic drive not in neutral.
Battery discharged or defective.
PTO drive engaged.
Defective safety switch.
Defective starter.
Defective solenoid.
Loose electrical connections.
Defective key switch.

Engine Cranks But Will Not Start

Empty fuel tank.
Restricted fuel tank vent.
Fuel shut-off valve closed.
Clogged or restricted fuel line.
Breaker points worn or pitted.
Defective spark plugs.
Battery not fully charged.
Loose electrical connections.
Faulty condenser.
Defective ignition coil.
Dirt in fuel system.

Engine Starts Hard

Defective spark plugs.
Breaker points worn or pitted.
Loose wires or connections.
Restricted fuel tank vent.
Clogged or restricted fuel line.
Broken choke or throttle cable.
Dirt or water in fuel system.
Carburetor not properly adjusted.
Wrong valve clearance.
Head gasket leaking.
Low compression.

Engine Starts But Fails To Keep Running

Restricted fuel tank vent.
Carburetor not properly adjusted.
Broken choke cable.
Dirt or water in fuel system.
Breaker points not properly adjusted.
Loose wires or connections.
Defective head gasket.
Faulty condenser.

Engine Runs But Misses

Breaker points not properly adjusted.
Defective spark plugs.
Loose wires or connections.
Carburetor float leaking or not properly adjusted.
Dirt or water in fuel system.
Wrong valve clearance.
Faulty coil.
Air intake or shrouding plugged.

Engine Misses Under Load

Defective spark plugs.
Carburetor not properly adjusted.
Incorrect spark plugs.
Breaker points not properly adjusted.
Ignition out of time.
Dirt or water in fuel system.

Engine Will Not Idle

Idle speed too slow.
Idle mixture needle not properly adjusted.
Dirt or water in fuel system.
Restricted fuel tank filler cap.
Defective spark plugs.
Wrong valve clearance.
Low engine compression.

Engine Loses Power

Crankcase low on oil.
Air intake or shrouding plugged.
Excessive engine load.
Restricted air filter.
Dirt or water in fuel system.
Carburetor not properly adjusted.
Defective spark plugs.
Too much oil in crankcase.
Low engine compression.
Worn cylinder bore.

Engine Overheats

Air intake or shrouding plugged.
Carburetor not properly adjusted.
Too much oil in crankcase.
Crankcase low on oil.
Excessive engine load.

Engine Knocks

Engine out of time.
Excessive engine load.
Crankcase low on oil.

Engine Uses Excessive Amount of Oil

Clogged breather assembly.
Breather not assembled properly.
Worn or broken piston rings.
Worn cylinder bores.
Wrong size piston rings.
Worn valve stems and/or valve guides.
Incorrect oil viscosity.

Fig. 9-3 Troubleshooting chart for the mechanic

- Do not operate an engine without all safety guards in place. Take care that your clothing, hands, feet, and hair do not get caught in any moving parts.

- Do not touch any electrical wires or electric components on a running engine. A faulty part or unguarded terminal represents an electrical shock hazard that could cause injury or death.

- If the engine has a battery system that must be disconnected, disconnect the negative or ground cable first. When reconnecting the battery, reconnect the positive cable first. Be careful not to spill any battery acid as it can cause serious burns.

A typical tune-up procedure includes the following steps. The mechanic's repair manual for the particular engine should be at hand.

1. Inspect the air cleaner then clean and reassemble it. Is the cleaner damaged in any way? Is the filter element too clogged for cleaning? If so, replace the cleaner and/or element. Clean the unit according to the manufacturer's instructions.

2. Clean the gas tank, fuel lines, and any fuel filters or screens.

3. Check the compression. This can be done by slowly turning the engine over by hand. As the piston nears top dead center, considerable resistance should be felt. As top dead center is passed, the piston should snap back down the cylinder. A more thorough test can be made with a compression

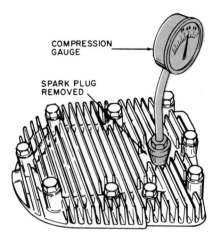

FIRST TEST: WITHOUT OIL IN CYLINDER
SECOND TEST: SQUIRT A FEW DROPS OF OIL ONTO PISTON THROUGH SPARK PLUG HOLE. TURN ENGINE OVER 6 TO 8 REVOLUTIONS TO GET OIL TRANSFERRED TO PISTON RING AREA, THEN MAKE COMPRESSION TEST.

Fig. 9-4 Checking compression with a gauge

gauge, figure 9-4. The manufacturer's repair manual should be consulted for normal and minimum acceptable compression pressure.

NOTE: Many engines have an automatic compression release mechanism which holds the intake valve slightly open for cranking the engine during starting. This allows the operator to start the engine with less pull. However, the automatic compression release makes checking the compression difficult.

4. Check the spark plug; clean, regap, or replace it. Remove the high-tension lead from the plug and hold it about 1/8 inch from the plug base, figure 9-5. Be sure to hold the lead on the insulation. Turn the engine over. If a good spark jumps to the plug, the magneto is providing sufficient spark. Now replace the high-tension lead and

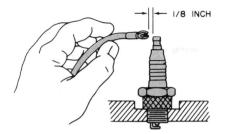

Fig. 9-5 Checking for spark

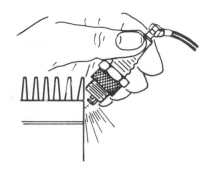

Fig. 9-6 Checking spark plug

lay the plug on a bare spot on the engine, figure 9-6. Turn the engine over. If a good spark jumps at the plug electrodes, the plug is good. This is not an absolute test but it is a good indication. If the plug is questionable, replace it.

5. Check the operation of the governor, figure 9-7. Be certain the governor linkages do not bind at any point.

6. Check the magneto. On most engines, the flywheel must be removed for access to the magneto parts. Use a flywheel or gear puller if available, figure 9-8. Otherwise, pop the flywheel loose by removing the flywheel nut and delivering a sharp hammer blow to a lead block held against the end of the crankshaft, figure 9-9.

Another method is to back off the flywheel nut until it is about 1/3 off the crankshaft and then strike it sharply. Do not hammer on the end of the crankshaft as this can damage the threads.

Adjust the breaker point gap and check the condenser and breaker point terminals for tightness, figure 9-10. Breaker points are set for 0.020 inch in most cases although the correct setting may vary according to

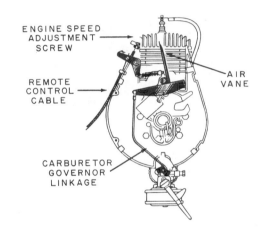

Fig. 9-7 Air vane governor parts

Fig. 9-8 Removing a flywheel using a special puller

Fig. 9-9 Removing the flywheel with a soft hammer

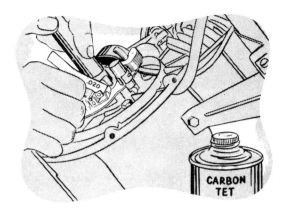

Fig. 9-10 Clean and check breaker points

the manufacturer. Turn the engine over until the breaker points reach their maximum opening, then check with a flat feeler gauge. Adjust the points if the setting is incorrect. Breaker points must line up, figure 9-11. Points can be cleaned if necessary. If the points are pitted or do not line up, they should be replaced, figure 9-12.

7. Fill the crankcase with clean oil of the correct type. (four-stroke cycle engines)

8. Fill the gasoline tank with regular gasoline. (four-stroke cycle engines)

   Fill the gasoline tank with the correct mixture of gasoline and oil. (two-stroke cycle engines)

9. Start the engine.

10. Adjust the carburetor for peak performance. See carburetor parts shown in figure 9-13.

RECONDITIONING

Reconditioning or overhauling an engine is generally the job for the mechanic — one who has the tools and know-how to do the job

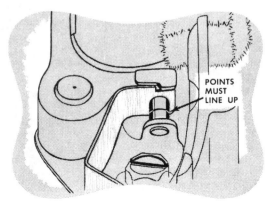

Fig. 9-11 Breaker points must line up

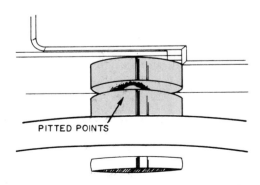

Fig. 9-12 Pitted breaker points should be replaced

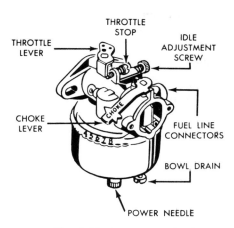

Fig. 9-13 Carburetor parts

correctly. However, much reconditioning and overhauling can be done by the amateur who has the basic tools and the correct approach. Care and precision are very important. Every part must be in place correctly, with none missing or left over.

Before beginning the disassembly of an engine, provide a spot to put the parts as they are removed. One good method is to lay out a large sheet of paper. As the parts are placed on the paper, label them so they will not be lost. It is often difficult to remember just where a part came from when forty or fifty pieces are laid out; also, it may be several days before the engine is reassembled.

CLINTON ENGINES TORQUE DATA – INCH POUNDS "Red Horse"					
		1600 A1600 A1690	1800 1890	2500 A2500	B2500 B2590 2790
Bearing Plate P.T.O.	Min. Max.	160 180	160 180	160 180	160 180
Back Plate to Block	Min. Max.	70 80	70 80	70 80	70 80

CLINTON ENGINES SERVICE CLEARANCES "Red Horse"		1600	A1600	A1690	1800	1890	2500	B2500	B2590	A2500	2790
Piston Skirt Clearance	Min. Max. Rework	.007 .009 .010	.007 .009 .010	.007 .009 .010	.007 .009 .010	.007 .009 .010	.0065 .0035 .010	.0065 .0085 .010	.0065 .0085 .010	.005 .007 .0085	.005 .007 .0085
Ring End Gap	Min. Max. Rework	.007 .017 .025	.007 .017 .025	.007 .017 .025	.007 .017 .025	.007 .017 .025	.010 .020 .028	.010 .020 .028	.010 .020 .028	.010 .020 .028	.010 .020 .028

CLINTON ENGINES TOLERANCES AND SPECIFICATIONS "Red Horse"	1600	A1600	A1690	1800	1890	2500	B2500	B2590	A2500	2790
Cylinder - Bore	2.8125 2.8135	2.8125 2.8135	2.8125 2.8135	2.9995 3.0005	2.9995 3.0005	3.1245 3.1255	3.1245 3.1255	3.1245 3.1255	3.1245 3.1255	3.1245 3.1255
Skirt - Diameter	2.8045 2.8055	2.8045 2.8055	2.8045 2.8055	2.9915 2.9925	2.9915 2.9925	3.117 3.118	3.117 3.118	3.117 3.118	3.1185 3.1195	3.1185 3.1195

Fig. 9-14 Torque data, clearances, tolerances, and specifications as excerpted from a service manual

A mechanic's handbook or service manual is essential for top-quality overhaul work. These books can sometimes be obtained directly from the engine manufacturer or borrowed from a mechanic. In only a few instances is this type of book supplied with the new engine. The service manual is quite detailed, fully explaining each step of overhaul or reconditioning. Allowances, clearances, torque data and other specifications are also given, figure 9-14. Without the use of the service manual there is too much guesswork involved. The amateur mechanic should have the service manual for the engine before any repair work is started.

Tools and Equipment

Unless the engine owner is a skilled mechanic and has an adequately equipped repair shop, the owner cannot perform all types of engine repair and overhaul. Most owners would not find it practical to purchase the equipment necessary to do all types of overhaul and repair. Professional repairers may have invested from five hundred to several thousand dollars in tools and equipment. Equipment may consist of: magneto testers, valve seat resurfacers, air compressors, steam cleaners, special sharpeners, grinders, lapping stands, special factory tools, general repair tools, repair parts, and others.

TORQUE SETTINGS – GENERAL

Tightening Torque Into Cast Iron or Steel

Size	Grade 2	Grade 5	Grade 8
8-32	20 in-lb	25 in-lb	
10-24	32 in-lb	40 in-lb	
10-32	32 in-lb	32 in-lb	
1/4-20	70 in-lb	115 in-lb	165 in-lb
1/4-28	85 in-lb	140 in-lb	200 in-lb
5/16-18	150 in-lb	250 in-lb	350 in-lb
5/16-24	165 in-lb	270 in-lb	30 ft-lb
3/8-16	260 in-lb	35 ft-lb	50 ft-lb
3/8-24	300 in-lb	40 ft-lb	60 ft-lb
7/16-14	35 ft-lb	55 ft-lb	80 ft-lb
7/16-20	45 ft-lb	75 ft-lb	105 ft-lb
1/2-13	50 ft-lb	80 ft-lb	115 ft-lb
1/2-20	70 ft-lb	105 ft-lb	165 ft-lb
9/16-12	75 ft-lb	125 ft-lb	175 ft-lb
9/16-18	100 ft-lb	165 ft-lb	230 ft-lb
5/8-11	110 ft-lb	180 ft-lb	260 ft-lb
5/8-18	140 ft-lb	230 ft-lb	330 ft-lb
3/4-10	150 ft-lb	245 ft-lb	350 ft-lb
3/4-16	200 ft-lb	325 ft-lb	470 ft-lb

Tightening Torque Into Aluminum

Size	Grade 2	Grade 5
8-32	20 in-lb	20 in-lb
10-24	32 in-lb	32 in-lb
1/4-20	70 in-lb	70 in-lb
5/16-18	150 in-lb	150 in-lb

CONVERSIONS

in-lb × 0.083 – ft-lb
ft-lb × 12 – in-lb
ft-lb × 0.1383 – kgm
ft-lb × 1.3558 – N m

Fig. 9-15 General torque settings for bolts

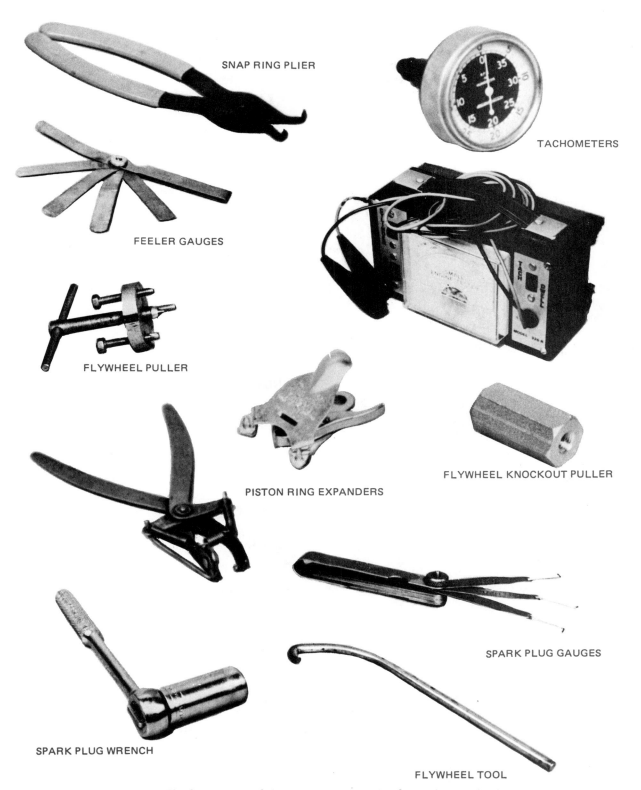

SNAP RING PLIER

TACHOMETERS

FEELER GAUGES

FLYWHEEL PULLER

PISTON RING EXPANDERS

FLYWHEEL KNOCKOUT PULLER

SPARK PLUG GAUGES

SPARK PLUG WRENCH

FLYWHEEL TOOL

Fig. 9-16 Some of the special tools needed for engine overhaul

A small investment in tools allows a person to perform many repair jobs. Most home workshops have some of the basic tools that are necessary. Several specialized tools can be added, figure 9-16. A tentative list of necessary tools includes the following:

- Screwdriver (various sizes)
- Combination wrenches (set)
- Adjustable wrench
- Socket set
- Torque wrench
- Deep well spark plug socket
- Needle nose pliers
- Feeler gauge
- Spark gap gauge
- Piston ring compressor
- Piston ring expander
- Valve spring compressor

Additional tools:
- Flywheel puller
- Compression gauge
- Tachometer
- Valve grinder – hand operated
- Arbor press

Basic Steps

Reconditioning or overhaul of an engine involves four basic steps: disassembly, inspection of parts, repair or replacement of worn or broken parts, and reassembly. In actual practice, a service manual would be followed for these steps. However, the following disassembly procedure is a general guide for a four-stroke cycle engine.

1. Disconnect the spark plug lead and remove the spark plug.

2. Drain the fuel system — tank, lines, carburetor.

3. Drain the oil from the crankcase.

4. Remove the air cleaner.

5. Remove the carburetor.

6. Remove the metal air shrouding, gas tank, and recoil starter.

7. Remove the flywheel.

8. Remove the breaker assembly and push rod.

9. Remove the magneto plate assembly.

10. Remove the breather plate assembly (valve spring cover).

11. Remove the cylinder head.

12. Remove the valves.

13. Remove the base.

14. Remove the piston assembly.

15. Remove the crankshaft.

16. Remove the camshaft and tappets.

17. Remove the mechanical governor.

As general information, the disassembly, inspection, repair, and reassembly for each engine part are discussed in this unit.

FLYWHEEL

Removal of the flywheel is necessary to gain access to the ignition system parts. If a special flywheel puller designed for the specific engine is available, use it. Otherwise use a flywheel puller. Be sure the flywheel nut or any threaded starter mechanism is removed before trying to pull the flywheel. Opposing pressure against the flywheel and the end of the crankshaft will break the contact of the tapered hole and the tapered shaft. Care should be taken with aluminum flywheels since excessive force can cause them to break.

If a flywheel puller is not available, the flywheel can be removed by first removing the flywheel nut and then striking the end of the crankshaft with a plastic, lead, or other soft-face hammer. A sharp blow on the shaft while exerting some outward pull on the flywheel will separate the flywheel from the shaft. Take care not to damage the end of the crankshaft, especially the threads. Engines that rotate clockwise have right-handed threads on the crankshaft and engines that rotate counterclockwise have left-handed threads.

The magnets in the flywheel can be damaged if the flywheel is dropped or if the magnets are hammered. Excessive heat causes the magnets to lose magnetism.

The key that aligns the crankshaft and flywheel is small; care should be taken not to lose it. If the key is damaged or worn, it should be replaced since poor alignment can affect engine timing. When reassembling the flywheel onto the crankshaft, carefully fit the key into the keyway, figure 9-17. Also, be

sure that no washers are forgotten when the flywheel is reinstalled.

MAGNETO PARTS

The magneto parts (high-tension coil, breaker points, condenser) may need to be removed for checking or replacement. The coil may be visually inspected for cracks and gouges in insulation, evidence of overheating, and the condition of the leads where they go into the coil. If an ignition coil tester is available, the coil can be checked for firing, leakage, secondary continuity, and primary continuity. This checking should be done using the procedure recommended by the manufacturers.

Breaker points that are corroded, pitted or dirty, provide a poor electrical contact. This lowers the magneto's ability to produce the proper voltage in the primary windings. Also, when the breaker points open, less voltage is induced because a clean, quick break is not produced. Breaker points that are in poor condition can seriously reduce the voltage available to the spark plug.

The setting of the breaker points (usually 0.020 inch) affects the engine timing. If the points are set for too small an opening, the spark is retarded. The spark is late because the cam must rotate further before the points open. Too wide an opening causes the spark to be advanced. Both conditions cause an engine to lose power. The two points should meet each other flat with their surfaces in full contact with each other. Points have a chrome or silvery appearance when new, and become gray as they are used.

The condenser can be visually inspected for dents, terminal lead damage and broken mounting clips. A condenser tester is used to check the condenser for capacity, leakage and

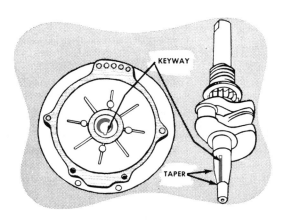

Fig. 9-17 Keyway on tapered end of the crankshaft and the flywheel

series resistance, figure 9-18. The test procedure suggested by the test equipment manufacturer should be followed.

Manufacturers often recommend replacing the condenser when the points are replaced even though the condenser is still reliable and in good condition. If one of the breaker points is a part of the condenser, the entire condenser must be replaced.

When the magneto is reassembled, the laminated iron core must be as close to the flywheel and its magnets as possible without touching or rubbing as the flywheel revolves. The closer the magnet is to the core (or pole shoes) the stronger the magnetic influence of the flywheel magnets. A space of about 0.007 inch to 0.012 inch is usually suggested, about the thickness of a postal card.

ENGINE TIMING

Engine timing refers to the magneto timing to the piston — the position of the piston just as the breaker points start to open. Timing is set at the factory but it is possible for timing to cause engine trouble. If the breaker points open too late in the cycle, power is lost. If the breaker points open too early in the cycle, detonation can result. Improper engine timing can be caused by the

- crankshaft and camshaft gears being installed one tooth off
- spark advance mechanism being stuck
- breaker points incorrectly set
- magneto assembly plate loose or slipped
- rotor incorrectly positioned.

These causes of trouble depend on the engine and the type of magneto.

The position of the piston for timing a magneto varies from engine to engine. The

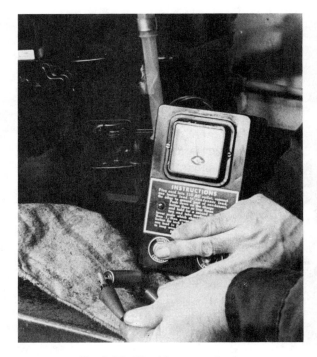

Fig. 9-18 Checking a condenser

piston can be at top dead center (TDC) or slightly before the top dead center (BTDC). Check engine specifications for this information.

One common method of checking timing is to locate TDC. This is the point where the piston does not seem to move as the flywheel is rotated. A dial indicator can be used for greater accuracy. With TDC located, place a reference mark on the flywheel and the magneto assembly plate, figure 9-19. Now remove the flywheel. Rotate the crankshaft until the breaker points barely open (0.001 inch). Carefully replace the flywheel. Put another mark on the magneto plate aligned with the flywheel reference mark. The difference between the marks on the magneto plate is the magneto timing to the piston. This can be figured in degrees by dividing the number of flywheel vanes into 360 degrees. (Twenty vanes; each vane 18

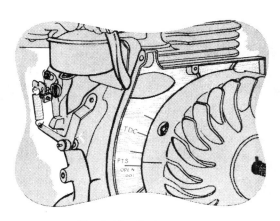

Fig. 9-19 Timing marks

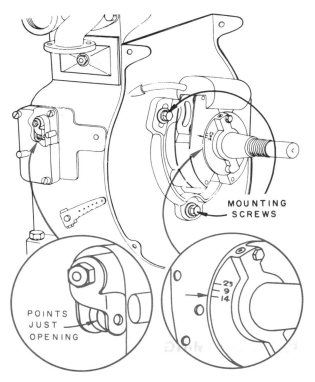

Fig. 9-20 Timing the rotor and armature to the piston. The numbers represent model numbers.

degrees.) The correct number of degrees of firing before top dead center is found in the engine specifications.

If the engine has a magnetic rotor magneto, the rotor and armature must be timed to the piston, figure 9-20. Timing is correctly set when the engine leaves the factory. However, if the armature has been removed or the crankshaft or cam gear replaced, it is necessary to retime the rotor. Breaker points are first set correctly (0.020 inch). The rotor is on the shaft correctly and tightened. The armature is mounted but mounting screws are not tight. Turn the crankshaft until the breaker points just start to open. (Place a piece of tissue paper between the points to detect when the points let go.) Now turn the armature slightly until the timing marks on the rotor and the armature line up.

Provisions are made on many engines to check the timing of the engine when the engine is running. This timing is done with a timing light, figure 9-21. Follow the instructions given for the particular timing light and engine. The timing light flashes each time the breaker points open. Reference marks on the engine must be located — one on a stationary part of the engine and one on the flywheel.

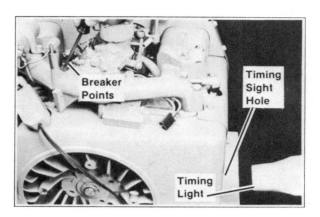

Fig. 9-21 Timing light method

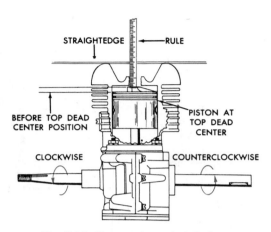

Fig. 9-22 Two-cycle engine timing

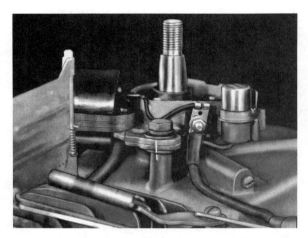

Fig. 9-23 Two-cycle timing marks

The timing light is directed at the flywheel. If the two marks line up, timing is correct. If the marks do not line up, the breaker plate must be loosened and shifted to a position where the marks do line up.

Two-cycle engine timing is quite similar, figure 9-22. One common procedure is to remove the spark plug and locate top dead center using a ruler and straightedge across the head. With TDC located, back the piston down the cylinder the correct distance (check manufacturer's specifications for the measurement before TDC). Now the breaker points should just begin to open. If timing is incorrect, loosen the stator plate setscrew and rotate the stator plate slightly until the points just begin to open. Retighten the stator plate setscrew. See figure 9-23 for an example of two-cycle timing marks.

CYLINDER HEAD

The cylinder head must be removed if work is to be done on the piston, piston rings, connecting rod, or valves. It must also be removed if the combustion chamber is to be

cleaned. The head bolts or nuts should be removed and set aside. The head gasket may be stuck to the head, cylinder, or both; in many cases the gasket is badly damaged while the head is being removed. If the gasket is damaged, it should be discarded.

Carbon deposits should be carefully scraped from the cylinder head and all old gasket material scraped away from the contact

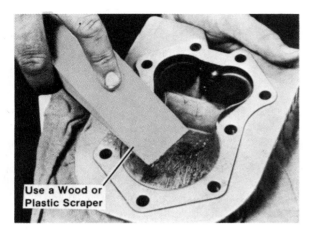

Fig. 9-24 Cleaning cylinder heads

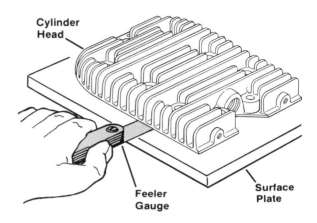

Fig. 9-25 Checking cylinder head flatness

surfaces, figure 9-24. The cylinder head can be checked for warpage by laying it on a *surface plate* (a large block of metal with a perfectly flat surface), figure 9-25. If the cylinder head is warped, it should be replaced.

A new head gasket should be used upon reassembly. The cylinder head bolts or nuts should be tightened in the correct sequence. To be certain of the correct degree of tightness, a torque wrench should be used (generally 14 to 18 ft-lb), figure 9-26.

VALVES

Valve problems are fairly common and can be quite serious. Faulty valves affect engine power and cause hard starting. A poorly seated valve is one of the possible causes for poor compression. If it is reasonably certain that the piston rings are in good shape and well oiled and the head gasket is air tight and a compression test indicates leakage, the leakage is probably caused by a valve problem. A quick test is to remove the spark plug and

hold a finger over the spark plug hole while turning over the engine. Compression should blow the finger away from the hole with noticeable force.

With the cylinder head removed the valves can be visually inspected. Look for problems such as carbon fouling, burned valves, valve corrosion, or valve sticking.

Fig. 9-26 Tightening the cylinder head bolts with a torque wrench

Fig. 9-27 Burned valve

Carbon Fouling.   Carbon deposits may have built up to the point where the valves do not fully seat. Also, carbon particles that have broken loose may be under the valve seat.

Burned Valves.   Burned valves are often caused by improper tappet clearance, figure 9-27. The valve may seat nicely when the engine is cold. However, when the parts heat up they expand and the valve cannot fully seat. Being constantly exposed to the hot gases of combustion, the valve may burn and the valve seating becomes progressively worse.

Valve Corrosion.   Valve corrosion can be seen when the valve is removed, figure 9-28. The top of the valve stem (neck) is worn away or thinned out. The problem is usually caused by having too lean a fuel mixture or by condensation. Condensation can form during storage or when the engine is frequently stopped before it reaches its normal operating temperature.

Valve Sticking.   A stuck valve means that the valve stem and valve guide have built up deposits of carbon and gum, figure 9-29. The deposits build up more rapidly in hot weather

Fig. 9-28 Valve corrosion

and under heavy engine loads. An engine that runs hot due to clogged radiating fins or a clogged air inlet screen builds up valve stem deposits faster than normal. Old gasoline often causes a rapid buildup of gum deposits. When the engine is cold, the valves may not stick; as the engine warms up the deposits

Fig. 9-29 Gum deposits cause valve sticking

expand and the valves stick. To correct the problem, the valve guides must be reamed out and the valves cleaned or replaced. Exhaust valves take more punishment than intake valves and are more subject to valve problems.

Engine owners may be able to do some valve work themselves. However, a large amount of work or a complete renewal of the valve system should be done by a mechanic who has the tools and necessary experience. A complete valve job includes:

- Installing new valves
- Installing new valve guides
- Installing new valve seats
- Installing new valve springs
- Grinding valve seats
- Lapping valves

One of the first check points is the *tappet clearance* (space between tappet and end of valve). This clearance is checked with a feeler gauge, figure 9-30. On most small engines the clearance can be enlarged by grinding a small amount from the end of the valve. If the clearance is already too large, the valve must be replaced. Some small engines have adjustable tappet clearances.

To remove the valves, first compress the valve spring; then slip off or slip out the valve spring retainers, sometimes called keepers. Pull the valve out of the engine. The valve spring and associated parts will also come out, figure 9-31.

With the valve out, the valve stem, face, and head can be closely inspected and cleaned. The valve guide and valve seat can also be inspected.

Many engines have replaceable valve guides. If these are worn or otherwise damaged, they must be pressed out using an arbor

Fig. 9-30 Checking tappet clearance

Fig. 9-31 Using a valve spring compressor to remove the valve springs

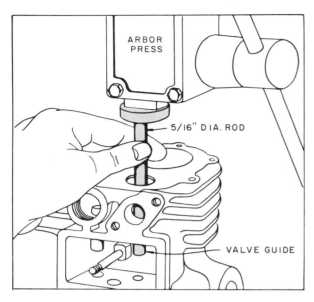

Fig. 9-32 Removing a worn valve guide

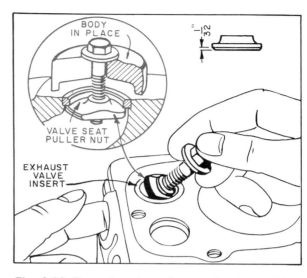

Fig. 9-33 Removing the exhaust valve seat with a special puller

press or carefully driven out with a special punch, figure 9-32. Also, the exhaust valve seat is removable on many engines. If the seat is beyond regrinding, remove it and reinstall a new valve seat. A special valve seat extracting tool is used for this job, figure 9-33.

If the valves and/or valve seats need regrinding, this can be done with special valve grinding equipment, figure 9-34. However, in some cases hand valve grinders can be used, figure 9-35.

If either or both of these parts have been replaced or reground, the parts must be

Fig. 9-34 Valve grinder

Fig. 9-35 Reconditioning a valve seat

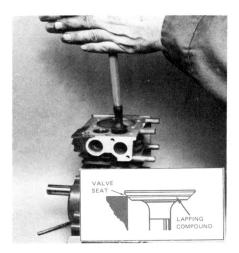

Fig. 9-36 Lapping valves

lapped to provide the perfect seal necessary for valve operation, figure 9-36. When lapping, use a small amount of lapping compound. Rotate the valve against the seat a few times until the compound produces a dull finish on the valve face. Do not lap the valves too heavily. When the valves are replaced, oil the stems and be sure the exhaust valve goes in the exhaust side and that the intake valve goes in the intake side.

PISTON AND ROD ASSEMBLY

In order to gain access to the piston and connecting rod assembly, the connecting rod bolts must be reached. On many engines the mechanic must first remove the engine from the base or sump, figure 9-37. This is done by loosening the bolts and breaking the seal. In most cases, a new gasket should be installed upon reassembly to prevent oil leakage. On engines that do not have a separate base, the bearing plate on the power takeoff side will need to be removed.

To remove the piston rod assembly, remove the connecting rod cap from around the

Fig. 9-37 Engine base, gasket, engine

crankshaft. Look for the boss marks that indicate how the end of the connecting rod and the rod cap go together. They must be reassembled in the same position. Push up on the bottom of the connecting rod and force the piston rod assembly up and out of the cylinder, figure 9-38. If there has been

Fig. 9-38 Connecting rod and piston removal

carbon buildup in the cylinder above the point of ring travel, the carbon should be scraped clean before the piston is removed. Mark the piston so it can be reinstalled the same way it came out.

CYLINDER

The cylinder should be checked for score marks. Scoring or wear scratches in the area of ring travel cause increased oil consumption and reduced engine power. Also the cylinder should be checked for distortion or warpage and size with a cylinder gauge, figure 9-39.

If the cylinder appears to be in good condition, it can be deglazed with a finish hone to prepare the cylinder for new rings, figure 9-40. However, a cylinder in poor condition may have to be rebored to a larger size which will also require a new and oversize piston for the cylinder.

PISTON

The piston should be checked for scuffing, scoring and scratches. Scoring are relatively deep marks that are caused by overheating and the metal fusing against the cylinder and wiping away metal. Lack of lubrication is usually the cause. Scuffing marks are not as deep. Scratches may be caused by abrasive materials entering the engine through the oil supply or air cleaner, figure 9-41.

The size of the piston itself can be checked with a micrometer. Generally it is 0.005 inch to 0.010 inch smaller than the cylinder. If the piston is too small, the piston skirts may slap the cylinder as it moves up and down.

PISTON PIN

The piston pin can be removed by dislodging the retainer rings from their grooves. Needle nose pliers work quite well for this operation. The piston pin is usually quite tight in the piston so it must be driven out with a soft punch and hammer. Reassemble the piston pin and connecting rod in the same positions.

PISTON RINGS

If the piston size and condition have checked out all right, the piston rings should be checked next. Check the edge gap with a

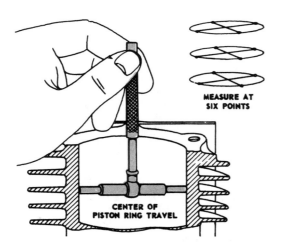

MEASURE AT SIX POINTS

CENTER OF PISTON RING TRAVEL

Fig. 9-39 Check cylinder bore

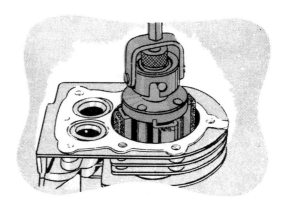

Fig. 9-40 Deglazing a cylinder with a finish hone

Stuck, Broken Rings

Abrasive Scratched Rings

Scored Piston and Rings

Fig. 9-41 Piston damage

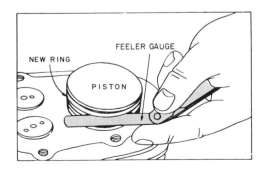

feeler gauge, with the rings still on the piston, figure 9-42. The correct edge gap clearance is found in the engine overhaul specifications. Remove the piston rings from the piston with a piston ring expander, figure 9-43. Carefully

Fig. 9-42 Checking the ring groove with a feeler gauge

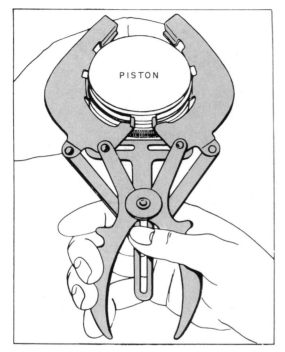

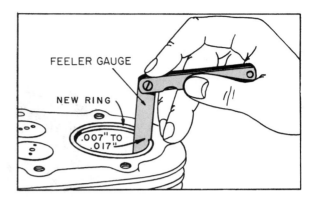

Fig. 9-44 Checking the ring gap with a feeler gauge (end gap)

Fig. 9-43 Piston ring tools

Fig. 9-45 Piston ring compressor

put the ring in the cylinder and check the end gap with a feeler gauge, figure 9-44. Too little end gap can cause the ring to freeze when it becomes hot and expands. Too much end gap may allow blow-by and a resulting loss of power. Piston ring grooves should be cleaned to remove any carbon accumulations.

New piston rings are usually installed during engine overhaul.

Use a piston ring compressor to reinstall the piston assembly, figure 9-45. Remember to place the piston in the cylinder the same way it came out. Also, put the connecting rod cap on the same way it came off, finding the match marks, figure 9-46. A torque wrench should be used to tighten the connect-

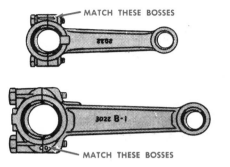

Fig. 9-46 Match boss marks when reassembling the connecting rod and connecting rod cap

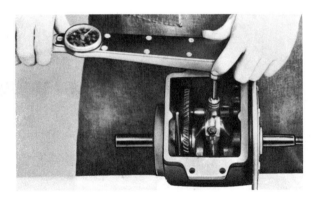

Fig. 9-47 Torque wrench used to correctly tighten connecting rod cap

ing rod cap bolts according to the manufacturer's specifications, figure 9-47.

CRANKSHAFT

The crankshaft should be removed and checked for scoring and any metallic pickup. The journals and crankpin should be checked with a micrometer for roundness. The gear and keyway should be checked for wear. Also, check the crankshaft for straightness, figure 9-48. In some cases, the main ball bearings may come out with the crankshaft. Upon reinstallation, the bearings may have to be pressed into place. An arbor press is good for this job, figure 9-49.

CAMSHAFT AND TAPPETS

The camshaft and valve tappets can also be removed for inspection and repair. The camshaft pin should be driven out with a drift punch from the power takeoff side of the engine. Upon reassembly of the camshaft and crankshaft be certain to line up the timing marks, figures 9-50 and 9-51.

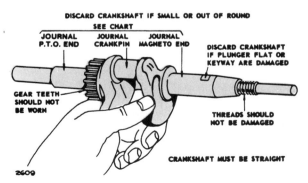

Fig. 9-48 Crankshaft check points

Fig. 9-49 Pressing in the main ball bearing

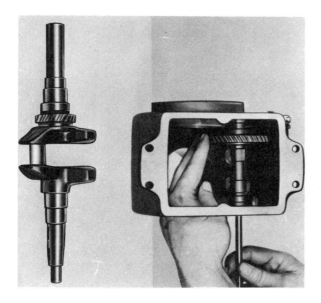

Fig. 9-50 Reinstalling the camshaft

Fig. 9-51 Reinstalling the crankshaft

REVIEW QUESTIONS

1. Define troubleshooting.

2. Of what value is a troubleshooting chart?

3. Can troubleshooting be done by an individual engine owner or operator?

4. Define engine tune-up.

5. What does engine reconditioning or overhaul involve?

6. Can any reconditioning or overhaul be accomplished by the amateur mechanic?

7. Why is the careful layout of parts important during disassembly?

8. Why is a mechanic's handbook or service manual essential for the reconditioning of an engine?

9. Explain engine timing.

10. What might a complete valve job include?

CLASS DISCUSSION TOPICS

- Discuss the troubleshooting chart.
- Discuss the steps in engine tune-up.
- Discuss the importance of care, accuracy, etc. in engine reconditioning and overhaul work.
- Discuss the tools necessary for reconditioning work.
- Discuss the steps in engine reconditioning.

CLASS DEMONSTRATION TOPICS

- Demonstrate troubleshooting by putting troubles into an engine and having students observe engine operation and then troubleshoot the engine. Simple examples: no fuel in tank, poor compression, shorting bar on plug, fuel shutoff valve closed, spark plug lead loose, bad spark plug, etc.
- Demonstrate how to tune an engine.
- Demonstrate the correct use of the basic tools used for engine repair.
- Demonstrate timing an engine.
- Demonstrate the tightening sequence and correct torque on a cylinder head.
- Demonstrate removing and inspecting valves for wear and damage.
- Demonstrate checking tappet clearance.
- Demonstrate lapping valves.
- Demonstrate removing and inspecting piston assembly for damage and wear.
- Demonstrate removal and inspection of the camshaft and tappets.

LABORATORY EXPERIENCE 9-1
CLEAN AND ADJUST THE BREAKER POINTS

The correct setting of the breaker points is essential to the operation of the magneto. The maximum opening of the points is most often 0.020 inch, however, some magnetos specify 0.018 inch, 0.023 inch or other settings; check the manufacturer's recommendations.

The breaker points should fully line up with each other and they should contact flat against each other. The breaker points have a chrome or silvery appearance when they are new. As the points are used they become gray. If they have become pitted they should be replaced.

STUDENT ASSIGNMENT

You are to clean and reset the breaker points. It is essential that you record your work in a work record box as you complete each step of the procedure.

Procedure

Your instructor may supplement or revise specific steps of the procedure which follows since there are many makes of engines. The following procedure is a general guide:

1. Remove the spark plug lead.

2. Remove air shroud, grass screens, etc. in the area of the magneto.

3. Remove the flywheel. This is not necessary if the points are not located under the flywheel.

4. Remove breaker point cover.

5. Rotate the crankshaft until maximum point opening is attained.

6. Check the breaker point gap using a flat feeler gauge prior to cleaning and resetting.

7. Clean the points with carbon tetrachloride on a lint-free cloth.

8. Loosen the breaker point assembly lock screw.

9. With the points at maximum opening, turn the breaker point adjusting screw to attain the correct setting. Check the setting with a flat feeler gauge.

10. Tighten the breaker point assembly lock screw.

11. Rotate the crankshaft several times.

12. Recheck the breaker point setting with a flat feeler gauge.

WORK RECORD BOX

PART	DISASSEMBLY (nuts, bolts, etc.)	OPERATION PERFORMED	TOOL USED

REVIEW QUESTIONS

Reset the breaker points, then answer the following questions.

1. What was the breaker point gap before adjustment?

2. What is the correct breaker point setting for the engine?

3. Were the breaker points under the flywheel? If so, could the breaker cam be seen?

4. Were the breaker points mounted outside the flywheel?

LABORATORY EXPERIENCE 9-2
TEST SPARK PLUG AND MAGNETO OUTPUT

Trouble in the ignition system may stem from the spark plug or from the magneto itself. It is often helpful in troubleshooting to determine (1) if the magneto is delivering high-tension voltage to the spark plug, and (2) if the spark plug is delivering a spark. Poor ignition can result in engine missing, poor performance under heavy load, and hard starting. If trouble exists, these two simple checks can help to establish whether or not the ignition system is the source of difficulty.

STUDENT ASSIGNMENT

You are to perform the tests listed below, testing (1) the magneto output and (2) the output at the spark plug. Record your work in a work record box.

Procedure

Your instructor may supplement or revise specific steps of the procedure which follows since there are many makes of engines. The following procedure is a general guide:

1. Test strength of magneto output.
   a. Remove high-tension lead from spark plug.
   b. Hold high-tension lead about 1/8 inch from spark plug base.
   c. Turn engine over as if it were being started.
   d. If a good spark jumps to the plug, the magneto output is satisfactory.

2. Test strength of spark at spark plug.
   a. Remove high-tension lead from spark plug.
   b. Remove spark plug from engine and replace high-tension lead.
   c. Lay spark plug on engine so the plug base touches bare metal.
   d. Turn the engine over as if it were being started.
   e. If a good spark jumps at the electrodes, the spark plug is good.

NOTE: This is not an absolute test since it is more difficult for a spark plug to fire under compression. If the plug is questionable, install a new one.

WORK RECORD BOX

PART	DISASSEMBLY (nuts, bolts, etc.)	OPERATION PERFORMED	TOOL USED

REVIEW QUESTIONS

Upon completion of the two tests, answer the following questions.

1. Is it possible for the magneto to be good and the spark plug to be bad?

2. Is it possible to get the correct spark at the plug if the magneto is bad?

3. Why is it important to tighten the spark plug securely upon reassembly?

4. Describe the sparks that were seen. (Were they fat blue, thin blue, or yellowish blue?)

5. If the spark plug fires correctly when it is out of the cylinder, will it always fire correctly when it is under compression?

LABORATORY EXPERIENCE 9-3
ADJUST ENGINE TIMING

Engine timing refers to the magneto timing to the piston: the position of the piston just as the breaker points start to open. Timing is set at the factory but it is possible for timing to cause engine trouble. If the breaker points open too late in the cycle, power is lost. If the breaker points open too early in the cycle, detonation may result. Improper engine timing could be caused by the crankshaft and camshaft gears being installed one tooth off, spark advance mechanism stuck, breaker points incorrectly set, magneto assembly plate loose or slipped, or rotor incorrectly positioned. These causes of trouble depend on the engine and type of magneto.

The position of the piston for timing a magneto varies from engine to engine. The piston may be at top dead center (TDC) or slightly before top dead center (BTDC). Check engine specifications for this information.

NOTE: It is recommended that the adjustment of engine timing be attempted only when the engine repair manual is available for reference. Specific instructions and procedures should be followed. It should be impressed on the student that the work cannot be done without specific engine data.

STUDENT ASSIGNMENT

You are to correctly adjust the engine timing on your assigned engine. Your instructor will provide a manufacturer's book which has specific instructions and procedures to follow. Record your work in a work record box.

REVIEW QUESTIONS

Upon completion of the assigned work, answer the following questions.

1. Explain how to locate top dead center (TDC).

2. Why are the breaker points timed to open at TDC or before TDC? Why not after TDC?

3. List the steps that were followed in adjusting the engine timing.

4. What problems can incorrect engine timing cause?

WORK RECORD BOX

PART	DISASSEMBLY (nuts, bolts, etc.)	OPERATION PERFORMED	TOOL USED

LABORATORY EXPERIENCE 9-4
REPLACE PISTON RINGS

The piston rings fit into the slotted grooves around the piston. They provide the seal between the moving piston and the cylinder wall. The rings actually exert a pressure on the cylinder wall. An oil film between the rings and the cylinder wall prevents excessive friction and provides a power seal. The two types of piston rings are (1) compression rings, and (2) oil control rings.

The piston rings are exposed to extremely hard usage. Over a long period of time it is quite possible that replacement with new rings may be needed. Worn piston rings can result in loss of power, excessive oil consumption, crankcase dilution, or hard starting. The replacement of piston rings is a common repair job and is generally done when the engine is overhauled. It might be noted that the condition of the cylinder wall also has a bearing on piston ring efficiency. Worn or warped cylinder walls cannot be corrected with new rings; honing or reboring of the cylinder would be necessary.

STUDENT ASSIGNMENT

You are to remove the piston assembly from the engine, remove the piston rings, inspect the condition of the piston, install new rings, and reassemble the piston assembly in the engine. If you are actually installing new rings, you should have access to the piston ring clearance specifications for your engine. These clearances are checked with a feeler gauge to determine whether or not they are within the tolerance limits. It is essential that you record your work in a work record box as you complete each step of the disassembly procedure.

Disassembly Procedure

Your instructor may supplement or revise specific steps of the procedure which follows since there are many makes of engines. The following procedure is your general guide for four-stroke cycle engines:

1. Remove the necessary parts to expose the crankshaft.

2. Remove the cylinder head.

3. Remove the connecting rod cap (note markings on the cap so it can be reassembled in the same manner).

4. Push the piston assembly up and out of the engine (mark the piston so that it can be reinstalled the same way it came out).

5. Inspect the condition of the piston.

6. Check the edge gap with a feeler gauge (rings are still on the piston).

7. Remove the piston rings with a piston ring expander.

8. Carefully put the ring in the cylinder and check the end gap with a feeler gauge.

9. Clean the piston ring grooves, removing any carbon accumulations.

10. Install new rings or reinstall old rings using a piston ring expander.

11. Push the piston assembly back into the cylinder with the aid of a piston ring compressor.

12. Reassemble piston assembly to crankshaft.

WORK RECORD BOX

PART	DISASSEMBLY (nuts, bolts, etc.)	OPERATION PERFORMED	TOOL USED

REVIEW QUESTIONS

Upon completion of the assigned work, answer the following questions.

1. What are the two types of piston rings? What is the main function of each?

2. Describe the condition of the piston.

3. What are the edge gaps? Are they within tolerances?

4. What are the end gaps? Are they within tolerances?

5. Describe the markings or indications on the connecting rod and connecting rod cap that indicate correct reassembly.

6. What are some common troubles that are caused by bad piston rings?

LABORATORY EXPERIENCE 9-5
LAP VALVES

The valves must seat perfectly to form a seal for both the compression and power stroke. Valves operate under very adverse conditions. The exhaust valve, in particular, is subjected to intense heat and rapid motion. If the valves or valve seats become worn, cracked, or warped, sluggish engine operation as well as other problems result. Upon overhauling an engine, the valve and valve seat condition are checked. Sometimes either or both the valve seats and valves are reground. If valves or valve seats have been reground or replaced, they should be lapped. Lapping forms the necessary perfect seal between the seat and the valve.

STUDENT ASSIGNMENT

You are to remove the valves from the engine, inspect the valves and valve seats, replace necessary parts, lap the valves, and reinstall the valves. It is essential that you record your work in a work record box as you complete each step of the disassembly procedure.

NOTE: If you are working on an instructional engine, the teacher may want you to assume that new parts have been installed even though they have not, lapping the valves for practice and experience.

Disassembly Procedure

Your instructor may supplement or revise specific steps of the procedure which follows since there are many kinds of engines. The following procedure is a general guide:

1. Remove the valve spring cover exposing the valve springs.

2. Remove the cylinder head.

3. Compress the valve spring with a valve spring compressor.

4. Remove the valve spring retainers by slipping or flipping them out.

5. Remove the valve spring, still compressed.

6. Pull the valve out of the engine.

7. Inspect the valve and valve seat.

8. Replace the valve and valve seat, or regrind the valve and valve seat only upon the instructor's request.

9. Lap the valves. Place a small amount of lapping compound on the valve face and rotate the valve against its seat a few times. Lap the valves until there is a thin, dull ring around the entire valve face. The ring indicates that the valve will seat well. Do not lap the valves more than is necessary.

10. Clean the valve and oil the valve stem.

WORK RECORD BOX

PART	DISASSEMBLY (nuts, bolts, etc.)	OPERATION PERFORMED	TOOL USED

Reassembly Procedure

Reverse the disassembly procedure. Remember to put the exhaust valve in the exhaust side and the intake valve in the intake side. On some engines, a magnetic valve retainer inserter is a great help. The retainers often come out easily but are difficult to reinstall.

REVIEW QUESTIONS

Upon completion of the assigned work, answer the following questions.

1. Describe the condition of the valves.

2. Describe the condition of the valve seats.

3. What is the purpose of lapping valves?

4. What type of trouble can bad valves cause?

5. How is the exhaust valve different from the intake valve? Are there any special markings?

6. Can the exhaust valve seat on the engine be replaced?

7. Can the intake valve seat on the engine be replaced?

LABORATORY EXPERIENCE 9-6
CHECK ENGINE COMPRESSION

Good compression is essential for optimum engine performance. Fuel mixture must be tightly compressed to insure proper ignition and maximum power. Poor compression can be caused by worn piston rings, bad valves, worn or warped cylinders, leakage through the head gasket, or leakage around the spark plug. Poor compression is a common trouble, especially with older engines that are in need of an overhauling.

If an engine starts with difficulty or lacks power and is sluggish, a trouble-shooter might suspect poor compression as the possible cause for these troubles. Checking the engine's compression is a part of most tune-up procedures. Generally, engine manufacturers specify their own recommended method for checking engine compression.

STUDENT ASSIGNMENT

You are to perform a compression check on an engine. It is essential that you record your work in a work record box as you complete each step of the procedure.

Procedure

Your instructor may supplement or revise specific steps of the procedures which follow since there are many makes of engines. The following procedure is a general guide:

1. Checking compression *without* a compression gauge.
   a. Remove the spark plug high-tension lead from the spark plug.
   b. Turn the engine over slowly by hand. As the piston reaches top dead center, considerable resistance should be felt. As top dead center is passed, the piston should snap back down the cylinder, indicating good compression.

2. Checking compression *with* a compression gauge.
   a. Remove the spark plug lead from the spark plug.
   b. Remove the spark plug.
   c. Carefully clean any dirt or foreign matter from around the spark plug hole.
   d. Hold the compression gauge tightly against the spark plug hole. (Some gauges screw into the hole.)
   e. Turn the engine over as if starting it.
   f. Read the compression gauge. Readings of 60 to 80 pounds per square inch (psi) generally indicate good compression. However, some engines may have a compression range of 110 to 120 psi. The exact data for the engine you are working on should be at hand.

WORK RECORD BOX

PART	DISASSEMBLY (nuts, bolts, etc.)	OPERATION PERFORMED	TOOL USED

REVIEW QUESTIONS

Upon completion of the compression check, answer the following questions.

1. Name some of the engine parts whose failure can cause poor compression.

2. What are the common troubles that are caused by poor compression?

3. Does correcting a condition of poor compression always require a major repair job or overhaul of the engine?

4. Name the manufacturer of the engine used in this experience. What type of compression check did the manufacturer recommend?

5. If a compression gauge was used, record the compression (in psi). Was this pressure within the manufacturer's acceptable range?

LABORATORY EXPERIENCE 9-7
TOLERANCES AND ENGINE MEASUREMENTS

The average engine owner often does not fully appreciate the precision workmanship that is involved in machining and assembling engine parts. The many fast moving parts of the engine which operate under heavy load make precision a necessity. Tolerances are very close, always within a few thousandths of an inch. Mechanics who repair and overhaul engines must duplicate the factory precision if their work is to be satisfactory. The poorly equipped mechanic or the mechanic with little regard for precision can do an engine a great disservice.

NOTE: The following measurements are simple, only a feeler gauge is necessary. For tightening machine bolts and nuts, a torque wrench is necessary. If instruments such as inside and outside micrometers and dial indicators are available, the unit can easily be expanded.

STUDENT ASSIGNMENT

You are to disassemble the engine in order to measure certain clearances. Do not disassemble the engine any more than necessary. Upon reassembly of the engine, you will tighten certain nuts and machine bolts to the proper degree with a torque wrench.

If possible, obtain a repair manual for the engine and check the clearances against the tolerances. Also, use the torque data for the engine if it is available. It is essential that you record your work in a work record box as you complete each step of the disassembly procedure.

GENERAL PROCEDURE

Your instructor may supplement or revise specific steps of the procedure which follows since there are many makes of engines. The following procedure is a general guide. Read the entire procedure carefully before you begin work.

Disassembly Procedure

Disassemble the engine. Instructors will give the necessary steps in the engine disassembly procedure.

Checking Procedure

Using a feeler gauge:

1. Check the camshaft end clearance.

2. Check the crankshaft end clearance.

3. Check the connecting rod — large end side clearance.

4. Check the valve clearance for both intake and exhaust valves (measured between tappet and valve stem).

5. Check the piston skirt clearance at thrust face.

6. Check the ring end gap clearance.

7. Check the ring edge gap clearance.

8. Check the breaker points at maximum opening.

WORK RECORD BOX

PART	DISASSEMBLY (nuts, bolts, etc.)	OPERATION PERFORMED	TOOL USED

Reassembly Procedure

Using a torque wrench:

1. Tighten the connecting rod cap screws to 140 in-lb

2. Tighten the cylinder head cap screw to 200 in-lb

3. Tighten the flywheel nut to 45 ft-lb

4. Tighten the spark plug to 27 ft-lb

   NOTE: Use the torque data for the particular engine if it is available.

REVIEW QUESTIONS

1. Why is precision workmanship necessary in servicing an engine?

2. Briefly define engine tolerance.

3. Briefly explain the use of the feeler gauge.

4. How does a torque wrench differ from other wrenches used by the mechanic?

5. Why is a torque wrench used instead of a regular wrench in tightening the various parts?

6. Explain the term ft-lb; in-lb.

LABORATORY EXPERIENCE 9-8
TUNE-UP FOR SMALL ENGINES

Engine tune-up does not involve major engine repair work, rather, it is a process of cleaning and adjusting the engine so that it will give top performance. Tune-up can be done by an experienced engine owner or it can be done by a mechanic. Tune-up is typically done by the mechanic when an owner brings a lawnmower in for a spring checkup prior to summer use.

STUDENT ASSIGNMENT

You are to perform the general steps in engine tune-up on your assigned engine. Work carefully. Bring the engine up to its peak of cleanliness and operating efficiency. It is essential that you record your work in a work record box as you complete each step of the procedure.

Procedure

Your instructor may supplement or revise specific steps of the procedure which follows since there are many makes of engines. The following procedure is a general guide.

1. Inspect air cleaner; clean and reassemble the air cleaner.
2. Clean the gas tank, fuel lines, and any fuel filters or screens.
3. Check compression.
4. Check spark plug: clean, regap or replace.
5. Check operation of the governor.
6. Check magneto.
7. Fill crankcase with clean oil of the correct type.
8. Fill gasoline tank with regular gasoline. Be sure to mix oil with the gasoline if it is a two-stroke cycle engine.
9. Start engine.
10. Adjust carburetor for peak performance.

WORK RECORD BOX

PART	DISASSEMBLY (nuts, bolts, etc.)	OPERATION PERFORMED	TOOL USED

REVIEW QUESTIONS

Upon completion of the assigned work, answer the following questions.

1. Explain the reason for engine tune-up.

2. Can engine tune-up be done by an engine owner at home?

3. If the engine fails to pass the compression check, what could be causing the trouble?

4. Explain the danger involved in operating a gasoline engine in a closed building.

5. Why is the spark plug on a single-cylinder engine of particular importance in engine tune-up?

6. List the tools that are used in tuning the engine.

UNIT 10

Horsepower, Specifications, and Buying Considerations

OBJECTIVES After completing this unit, the student should be able to:
- **Discuss four kinds of horsepower.**
- **Explain the terms engine torque, volumetric efficiency, compression ratio, and piston displacement.**
- **List at least five things to consider when buying an engine.**

Horsepower (hp) is the yardstick of the engine's power, its capacity to do work. James Watt, the inventor of the steam engine, devised the unit. He assumed that the average horse could raise 33,000 pounds one foot in one minute. An engine that can lift 33,000 pounds one foot in one minute is a one-horsepower engine. The engine can lift 8250 pounds four feet in one minute, and still be delivering 1 horsepower.

As discussed in unit 1, the formula for horsepower is:

$$\text{Horsepower} = \frac{\text{Work}}{\text{Time (in minutes)} \times 33,000}$$

Remember that work is the energy required to move a weight through a distance. For example, lifting one pound, one foot, is one foot-pound of work. Lifting 50 pounds, two feet, is 100 foot-pounds of work. The element of time is not a factor.

In horsepower the time factor is added. For example, a boat with an outboard motor might cross a river in ten minutes. An identical boat with a larger motor might make the crossing in five minutes. The same amount of work has been accomplished by each engine but the larger horsepower engine did the work faster.

There are a variety of terms related to horsepower, such as brake horsepower, frictional horsepower, indicated horsepower, SAE or taxable horsepower, rated horsepower, developed horsepower, maximum horsepower, continuous horsepower, corrected horsepower, observed horsepower, and others. Several of these are discussed in the following paragraphs.

Brake horsepower is usually used by manufacturers to advertise their engine's power. Brake horsepower is measured either with a prony brake or with a dynamometer. The *prony brake* consists basically of a flywheel pully, adjustable brake band, lever, and scale measuring device. With the engine operating, the brake band is tightened on the flywheel and the pressure or torque is transmitted to the scale. The readings on the scale and other data are used to calculate the horsepower.

The *dynamometer* is a device that measures power produced by an internal combustion engine. The electric dynamometer contains a dynamo. As the engine drives the dynamo the current output can be carefully recorded. The more powerful the engine, the more current is produced. This current is correlated to horsepower.

Frictional horsepower is the power that is used to overcome the friction in the engine itself. The parts themselves absorb a certain amount of power. Losses included are mainly due to the friction of the piston moving in the cylinder but also included are losses due to oil and coolant pumps, valves, fans, ignition parts, and other accessories. These losses can amount to 10 percent of the brake horsepower.

Indicated horsepower is the power that is actually produced by the burning gases within the engine. It does not take into account the power that is absorbed by or used to move the engine parts. It is the sum of the brake horsepower plus the power used to drive the engine (frictional horsepower).

SAE or taxable horsepower is the horsepower used to compute the license fee for automobiles in some states.

$$\text{SAE hp} = \frac{D^2 \times \text{No. of Cylinders}}{2.5}$$

D = Diameter of bore in inches

Example: $\frac{(3.5 \times 3.5) \times 6}{2.5}$ = 29.4 hp

FACTORS AFFECTING HORSEPOWER

Many factors affect the horsepower of an engine. Some of these factors are discussed in the following paragraphs.

Engine torque is a factor that relates to horsepower. *Torque* is the twisting force of the engine's crankshaft. Torque can be compared to the force a person uses to tighten a nut. At first a small amount of torque is used, but more and more torque is applied as the nut tightens. Even when the nut stops turning, the person still may be applying torque. Motion is not necessary to have torque. Torque is measured in foot-pounds or inch-pounds. On an engine, the maximum torque is developed at speeds below the maximum engine speed. At top speeds, frictional horsepower is greater and volumetric efficiency is less.

Volumetric efficiency relates to horsepower also. It refers to the engine's ability to take in a full charge of fuel mixture in the short time allowed for intake. Engine design largely determines the volumetric efficiency of the engine. At high speeds the volumetric efficiency drops off. It can be increased with superchargers and turbochargers that force or blow air into the intake manifold. Also, volumetric efficiency can be increased by using multiple-barrel carburetors (two-barrel and four-barrel) with the extra barrels opening up at high speed when the demand for air is greater.

Compression ratio is another factor that affects horsepower. This is the relationship of the volume of the cylinder when the piston is at the bottom of its stroke compared to the volume of the cylinder (and combustion chamber) when the piston is at top dead center. For small gasoline engines it may be 6:1; for automobile engines, the compression ratio may be 8:1 or higher. The higher the compression ratio, the greater the horsepower delivered when the fuel mixture is ignited.

Piston displacement is the volume of air the pistons displace from the bottom of their stroke to top dead center. Piston displacement also relates to horsepower. Generally, the greater the piston displacement the greater the horsepower. Most engines deliver 1/2 to 7/8 horsepower per cubic inch displacement. High performance engines develop about 1 horsepower per cubic inch displacement. Supercharged engines can deliver much

more than 1 horsepower per cubic inch displacement. Displacement equals *area of bore × stroke × number of cylinders*, and is expressed in cubic inches.

Estimating Horsepower

The following formula can be used to estimate the maximum horsepower of a four-stroke cycle engine.

$$\text{Horsepower} = \frac{D^2 \times N \times S \times \text{rpm}}{11,000}$$

D^2 (Bore in Inches)

N (Number of Cylinders)

S (Stroke in Inches)

11,000 (Experimental Constant)

For a two-cycle engine, the formula can be used with 9000 as the experimental constant.

SPECIFICATIONS

Specifications for several models of some of the leading manufacturers of engines are listed below. In considering the best engine for a certain job, such information should be studied.

CUSHMAN MOTORS, Lincoln, Nebraska — Principal Usage: Utility Vehicles*

Model*	Rated HP	RPM	Bore	Stroke	Disp. Cu. in.	Comp. Ratio	2 or 4 Cycle	Weight
100	9		3.5	2.25	21.58	6.85/1	4	
200	18		3.5	2.25	43.16	6.85/1	4	

*Partial List of Motors

KOHLER CO., Kohler, Wisconsin — Principal Usage: All engine powered equipment

Model*	Rated HP	RPM	Bore	Stroke	Disp. Cu. in.	Comp. Ratio	2 or 4 Cycle	Weight
K91	4.1	4000	2 3/8	2	8.86	6.5/1	4	41
K161	7.0	3600	2 7/8	2 1/2	16.22	6.25/1	4	65
KV161	7.0	3600	2 7/8	2 1/2	16.22	6.25/1	4	65
L160	6.5	3600	2 7/8	2 1/2	16.22	6.25/1	4	106
K241	9.5	3600	3 1/4	2 7/8	23.9	6.00/1	4	105
K331	12.5	3200	3 5/8	3 1/4	33.6	6.25/1	4	173
K662	24.0	3200	3 5/8	3 1/4	67.2	6.00/1	4	250

*Other models too numerous to mention

JACOBSEN MANUFACTURING CO., Racine, Wisconsin — Principal Usage: Lawn Equipment

Model	Rated HP	RPM	Bore	Stroke	Disp. Cu. in.	Comp. Ratio	2 or 4 Cycle	Weight
J-125-H	2.25		2.0	1.5	4.71	5.5/1	2	
J-100-H	1.8		2.0	1.5	4.71	5.5/1	2	
J-125-V	2.25		2.0	1.5	4.71	5.5/1	2	
J-175-H	3.0		2.12	1.75	6.2	5.5/1	2	
J-175-V								
J-225-V	4.0		2.25	2.0	7.95	5.3/1	2	
J-321-V	3.0		2.12	1.75	6.2	5.0/1	2	
J-321-H								

BRIGGS AND STRATTON CORP., Milwaukee, Wisconsin — Principal Usage: General Power Use

Model*	Rated HP	RPM	Bore	Stroke	Disp. Cu. In.	Comp. Ratio	2 or 4 Cycle	Weight
92500	3	3600	2 9/16	1 3/4	9.02		4	19.5
100900	4	3600	2 1/2	2 1/8	10.43		4	30.5
130900	5	3600	2 9/16	2 7/16	12.57		4	30.75
60100	2	3600	2 3/8	1 1/2	6.65		4	22.25
80100	2.5	3600	2 3/8	1 3/4	7.75		4	22.25
80300	3	3600	2 3/8	1 3/4	7.75		4	25.25
190400	8	3600	3	2 3/4	19.44		4	45

*Other models too numerous to mention

CLINTON ENGINES CORP., Maquoketa, Iowa – Principal Usage: General Power Use

Model*	Rated HP	RPM	Bore	Stroke	Disp. Cu. In.	Comp. Ratio	2 or 4 Cycle	Weight
A2100	2.25	3600	2 3/8	1 5/8	7.2		4	21 1/2
100	2.50	3600	2 3/8	1 5/8	7.2		4	23
4100	2.75	3600	2 3/8	1 7/8	8.3		4	21 1/2
3100	3.00	3600	2 3/8	1 7/8	8.3		4	23
V1000	3.25	3600	2 3/8	1 7/8	8.3		4	36
V1100	3.75	3600	2 3/8	2 1/8	9.5		4	36
B1290	4.00	3600	2 15/32	2 1/8	10.2		4	45
V1200	4.50	3600	2 15/32	2 1/8	10.2		4	40
A1600	6.30	3600	2 13/16	2 5/8	16.3		4	87
B2500	9.60	3600	3 1/8	3 1/4	25.0		4	103
2790	10.30	3600	3 1/8	3 1/4	25.0		4	103

*Other models too numerous to mention

LAWN BOY, Galesburg, Illinois – Principal Usage: Lawn Mowers

Model*	Rated HP	RPM	Bore	Stroke	Disp. Cu. In.	Comp. Ratio	2 or 4 Cycle	Weight
C-10		4000	1 15/16	1 1/2	4.43	6.5/1	2	
C-12AA		4000	2 1/8	1 1/2	5.22	6.5/1	2	
C-18			2 3/8	1 1/2	6.65		2	
D-400			2 3/8	1 1/2	6.65		2	

*Other models too numerous to mention

WISCONSIN MOTOR CORP., Milwaukee, Wisconsin – Principal Usage: Heavy Duty Industrial

Model*	Rated HP	RPM	Bore	Stroke	Disp. Cu. In.	Comp. Ratio	2 or 4 Cycle	Weight
ACN	6	3600	2 5/8	2 3/4	14.88		4	76
BKN	7	3600	2 7/8	2 3/4	17.8		4	76
AENL	9.2	3600	3	3 1/4	23		4	110
AEH	7.4	3200	3	3 1/4	23		4	130
AGND	12.5	3200	3 1/2	4	38.5		4	180
THD	18	3200	3 1/4	3 1/4	53.9		4	220
VE4	21.5	2400	3	3 1/4	91.9		4	295
VF4	25	2400	3 1/4	3 1/4	107.7		4	295
VH4	30	2800	3 1/4	3 1/4	107.7		4	310
VG4D	37	2400	3 1/2	4	154		4	410
VR4D	56.5	2200	4 1/4	4 1/2	255		4	775

*Other models too numerous to mention

GRAVELY TRACTORS, INC., Dunbar, West Virginia — Principal Usage: Gravely Tractors

Model	Rated HP	RPM	Bore	Stroke	Disp. Cu. In.	Comp. Ratio	2 or 4 Cycle	Weight
L	6.6	2600	3 1/4	3 1/2	29.0	5/1	4	296

O & R ENGINES, Los Angeles, California — Principal Usage: General Power

Model	Rated HP	RPM	Bore	Stroke	Disp. Cu. In.	Comp. Ratio	2 or 4 Cycle	Weight
13B	1.0	6200	1.25	1.096	1.34	9/1	2	3.9
13A	1.0	6600	1.25	1.096	1.34	9/1	2	3.9
20A	1.6	7200	1.437	1.250	2		2	4.9

CHRYSLER OUTBOARD CORP. (Formerly West Bend) Hartford, Wisconsin
Principal Usage: Chain Saws, Scooters, Carts

Model	Rated HP	RPM	Bore	Stroke	Disp. Cu. In.	Comp. Ratio	2 or 4 Cycle	Weight
27824	3	4500	2	1 5/8	5.1	5.5/1	2	13 1/2
27825	3	4500	2	1 5/8	5.1	5.5/1	2	13 1/2
27852	3	4500	2	1 5/8	5.1	5.5/1	2	13 1/2
27854	3	4500	2	1 5/8	5.1	5.5/1	2	13 1/2
27612	5	5500	2 1/4	1 3/4	7.0	5.7/1	2	13 1/2
2760	5	5500	2 1/4	1 3/4	7.0	5.7/1	2	13 1/2

D. W. ONAN AND SONS, INC., Minneapolis, Minnesota
Principal Usage: Compressors, Truck Ref. Mowers, Go-Carts, Scooters, etc.
(Onan Generator units not listed)

Model	Rated HP	RPM	Bore	Stroke	Disp. Cu. In.	Comp. Ratio	2 or 4 Cycle	Weight
AS	5.5	3600	2 3/4	2 1/2	14.9	6.25/1	4	85
CCK	12.9	2700	3 1/4	3	50.0	5.50/1	4	148

TECUMSEH PRODUCTS CO., Grafton, Wisconsin — Principal Usage: General Power Use

Model*	Rated HP	RPM	Bore	Stroke	Disp. Cu. In.	Comp. Ratio	2 or 4 Cycle	Weight
AH47	3.2	4800	2	1 1/2	4.7		2	13 3/4
AH81	5.5	5000	2 1/2	1 5/8	7.98		2	13 3/4
AV47	2.2	3800	2	1 1/2	4.7		2	14 1/2
V55	5.5	3600	2 5/8	2 1/4	13.53		4	36 1/2
HR30	3.0	3600	2 5/16	1 13/16	7.61		4	28 1/4
HB30	3.0	3600	2 5/16	1 13/16	7.61		4	24

*Other models too numerous to mention

MCCULLOCH CORPORATION, Los Angeles, California — Principal Usage: Chain Saws

Model*	Rated HP	RPM	Bore	Stroke	Disp. Cu. In.	Comp. Ratio	2 or 4 Cycle	Weight
1-40		6000	2 1/8	1 3/8	4.9	5.5/1	2	18
1-50		6000	2 1/8	1 3/8	4.9	5.5/1	2	18
1-70		7000	2 1/4	1 1/2	5.3	7.0/1	2	21
1-80		7000	2 1/8	1 1/2	5.3	7.5/1	2	25

*Other models too numerous to mention

BUYING CONSIDERATIONS

The person who is shopping for a gasoline engine will have no trouble finding an engine for the job. There are many manufacturers of small gasoline engines who produce engines in a variety of horsepower that are designed and engineered to satisfy every need. It is not uncommon to find a manufacturer's basic model with variations to adapt the engine for many different jobs. Manufacturers of high-quality engines conduct many tests on their engines to assure quality performance, figures 10-1 to 10-3. Several points to consider in buying an engine are: cost, reputation of manufacturer, availability of repair parts, power requirements, engine warranties.

Cost. Beware of bargains. A good bargain can be found now and then but the general rule, "You get what you pay for," is a good one. This does not mean that the most

Fig. 10-1 Manufacturers of high-quality engines maintain elaborate testing laboratories.

Fig. 10-2 Cold room freezing test insures dependable operation in cold weather.

expensive engine is the best buy. The engine must be the right one for the intended usage. For cutting the grass once a week, an economical, dependable engine should be enough. However, if the engine will be exposed to continuous heavy usage it would be wise to buy a heavy-duty engine, one made to operate for a long period of time without failure. The additional cost is worthwhile in this case.

Reputation of the Manufacturer. Be certain the manufacturer's past record in the field justifies faith in the new engines. Talk with engine dealers and individual owners of engines. Their opinions and experience can help in making the best selection.

Availability of Repair Parts. If an engine breaks down and requires a new part, will it be readily available? Can the part be found at a local dealer or will it be necessary to send to a factory that may be many miles away? Delays caused by breakdowns can be costly as well as annoying. The availability of service and repair parts is an important consideration.

Power Requirements. Be certain the engine is big enough for the job. Constant overloading or constant operating at full throttle shortens engine life. A dealer can assist in finding the correct power plant for the job.

Manufacturers today produce engines that operate on either the four-cycle or the two-

Fig. 10-3 Endurance tests of engines insure their dependability.

cycle principle; both have their place and both are entirely successful. Some engines are water cooled, some are air cooled; some have simple splash lubrication, some have pump lubrication; some have fuel pumps, and some do not. In fact, no two engines look exactly alike or operate exactly alike.

Engine Warranties. Most engine manufacturers give a warranty to the purchaser of a new engine. A common warranty is good for ninety days. If during the ninety days after purchase any parts fail due to defective material or workmanship, the manufacturer will repair or replace the defective part.

In most cases the part or engine must be returned to the factory or to an authorized distributor. Most manufacturers require the owner to pay shipping charges to the factory. In some cases, the owner must also pay any labor costs involved in the repair.

If the engine has been damaged through misuse, negligence, or accident, most warranties are void. A few manufacturers require that the engine be registered shortly after purchase; if the engine is not registered, the warranty is not valid. Generally, engine components such as magnetos, carburetors, starters, etc. are only covered by the terms of their individual manufacturer.

REVIEW QUESTIONS

1. If an engine can lift 150,000 pounds a distance of one foot in three minutes, what is its horsepower?

2. In general, can larger horsepower engines do work faster than smaller horsepower engines?

3. What type of horsepower rating do most manufacturers use in advertising their engines?

4. What is a dynamometer?

5. In what way is SAE horsepower used?

6. Estimate the horsepower of a two-stroke cycle having a bore of 2 inches, one cylinder, a stroke of 1 1/2 inches, and operating at 3600 revolutions per minute.

7. What is engine torque?

8. What is volumetric efficiency?

9. What is the relationship between compression ratio and horsepower?

10. What is the relationship between piston displacement and horsepower?

11. List four important considerations for a person to think about before purchasing an engine.

CLASS DISCUSSION TOPICS

- Discuss how horsepower increases as rpm increases. (Remember that engines deliver much less than their rated horsepower at slow speeds.)

- Discuss and estimate the horsepower of a live engine in class. How does estimated horsepower compare with rated horsepower?

- Discuss how engine torque, volumetric efficiency, compression ratio, and piston displacement are functions of engine design.

- Discuss why there is a limit to the compression ratio that can practically be built into engines.

- Discuss the buying considerations for engines.

UNIT 11

Mechanical Transmission of Power

OBJECTIVES After completing this unit, the student should be able to:
- Explain the five simple machines.
- Recognize the three classes of levers.
- Calculate mechanical advantages of force, speed, and distance.
- Describe the power train and explain its function.

The power of the engine must be transmitted to the machines. In some cases, the machine can be fastened directly to the prime mover. This is desirable and simple. Few power losses occur this way. When the drive shaft of the prime mover is directly connected to the driven shaft of the machine or device, as, for example, a blade fastened to the engine's crankshaft on a rotary lawn mower, an ideal situation exists. Also, direct drive from the shaft can be seen in many machines that are driven by electric motors, such as grinders and polishers. The simple, ideal situation for power transmission can be found, but in most cases the machine application calls for something more complex.

Direct transmission of power may not cover all of the operating requirements of the given machine or mechanism. Factors such as speed changes, disengaging the machine from the prime mover, torque requirements, changes in direction of rotation, transmitting energy over long distances, transmitting energy around corners, converting rotary motion to linear motion, etc., cannot be satisfied by direct, positive linkage with the prime mover.

There are three primary methods of delivering the power of the prime mover to the machine that is to do the work: mechanical transmission, fluid transmission and electrical transmission.

Each of the three methods has its advantages and disadvantages. Engineers select the transmission system that fits the requirements. Often they use the systems in combination with each other. For example, coal is used to produce heat which converts water into steam. The steam is directed against the blades of the steam turbine. The steam turbine drives a generator producing electrical energy which is transmitted throughout the countryside by the use of wires. The energy may arrive at a factory where it is used to drive various electric motors.

The power produced by the electric motors may be transmitted to the machine through gears, belts, or other mechanical arrangements. The power of the electric motor may also be transmitted to a fluid power system which, in turn, transmits the power to the machine. There is an infinite number of combinations. This section deals with the physical principles and some of the techniques involved in mechanical transmission of power.

MECHANICAL TRANSMISSION OF POWER

Mechanical transmission principles enable a person to do more with the power that is available. The principles do not get any more out of the engine or other force-producing device. The total power being produced cannot be multiplied. However, the mechanical principles can adapt the power to a particular machine.

Several mechanical principles have been applied for centuries, long before engines were a source of power. The principles work well whether the source of power is an engine, a draft animal, a human, or a natural force. The principles discussed in this unit are often referred to as simple machines. These principles are restated by engineers in thousands of variations as seen in transmission systems and the machines themselves. It is sometimes difficult to draw a hard and fast line between

the transmission systems and the machine itself. The two may often blend into an integral unit. Countless applications of the simple machine principles are seen but basically the principles can change the:

- size of the applied force

- direction of the applied force

- speed that results from the applied force

These simple machines include the lever, wheel and axle, inclined plane and pulley.

Lever

Levers allow a person to lift, move, or hold far more than is possible without them. They are commonly seen. A teetertotter enables a small child to lift a man off the ground; a man uses a lever to move a boulder; a boy uses a pair of pliers to hold a bolt. Any long shaft or board will do for a lever as long as it has a pivot point, or fulcrum, to work against. In gaining a mechanical advantage with a lever, less force is used but it must apply over a greater distance to overcome and move the opposing object which resists movement.

Mechanical advantage is the number of times that force is multiplied. What is gained in applied force is lost in having to apply the force over a greater distance. The principle of the lever does not allow both a larger force and smaller distances moved. One is sacrificed for the other. With the lever shown in figure 11-1, an effort or force of 25 pounds (50.5 newtons) can lift 100 pounds (202 newtons), a mechanical advantage of four.

Levers fall into three categories: first class, second class, and third class levers.

First class levers place the fulcrum between the resistance and the effort force. These levers multiply force with a sacrifice in the

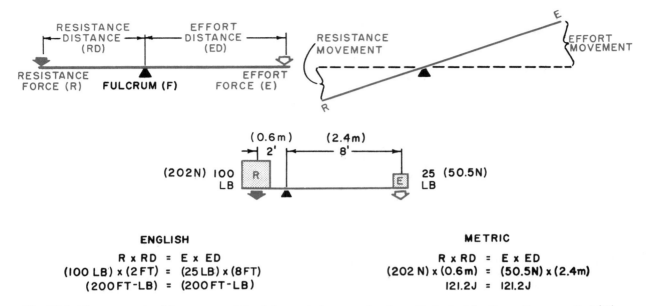

ENGLISH

R x RD = E x ED
(100 LB) x (2 FT) = (25 LB) x (8 FT)
(200 FT-LB) = (200 FT-LB)

METRIC

R x RD = E x ED
(202 N) x (0.6 m) = (50.5 N) x (2.4 m)
121.2 J = 121.2 J

Fig. 11-1   The moments of force around the fulcrum balance each other. Note that for larger force applied (R) or for greater distance applied (RD), the effort distance (ED) must be increased, or increase effort force (E).

distance the object is moved. The direction of movement is reversed, figure 11-2.

Second class levers place the fulcrum at the end of the lever with the resisting weight at a point along the lever. These levers also multiply force with a sacrifice in the distance the object is moved. The direction of object movement is the same as that of the applied force, figure 11-3.

Third class levers place the fulcrum at the end of the lever with the resisting weight at the opposite end of the lever. The applied force or effort is at a point along the lever. These levers increase effort by multiplying distance moved by the object and, therefore, have a negative mechanical advantage. A fishing rod illustrates this principle. Sometimes the fisherman is mystified at how a

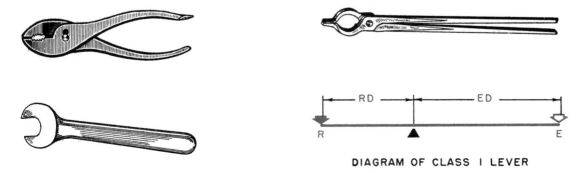

DIAGRAM OF CLASS I LEVER

Fig. 11-2   Examples of first class levers

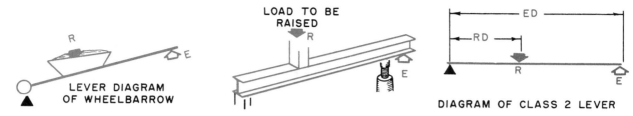

Fig. 11-3 Examples of second class levers

small fish can create such a large pull on his rod. The fish has the mechanical advantage, figure 11-4.

Wheel and Axle

Essentially the wheel and axle consists of two wheels firmly held together along the same axis. The axle may be short or it may be a very long shaft. Mechanical advantage is gained when the large wheel is moved through a distance. The axle moves also; its motion is less since its diameter is less, but the force twist, or torque, on the axle is multiplied, figure 11-5. A boat trailer winch is an example of this principle. The mechanical advantage of the winch allows a man of normal size to properly load and secure a heavy boat onto a trailer.

Study figure 11-6. Note that the mechanical advantage possible with the wheel and axle is essentially that of a Class 1 lever. The fulcrum is at the center of the wheel and the axle. The resistance distance (RD) is measured from the fulcrum to the perimeter of the axle. The effort distance (ED) is measured from the fulcrum to the perimeter of the wheel. In the perfect wheel and axle, these moments of force are equal: $E \times ED = R \times RD$.

Figure 11-7 shows an effort force of 10 pounds (44.5 newtons) applied to a 5-inch (127 millimeters) radius wheel. The resistance force acting on the axle of 1-inch (25.4 millimeters) radius which the effort force overcomes is found by the moment formula:

English

$$E \times ED = R \times RD$$
$$10 \text{ pounds} \times 5 \text{ inches} = R \times 1 \text{ inch}$$
$$\frac{50 \text{ pound-inches}}{1 \text{ inch}} = R$$
$$50 \text{ pounds} = R$$

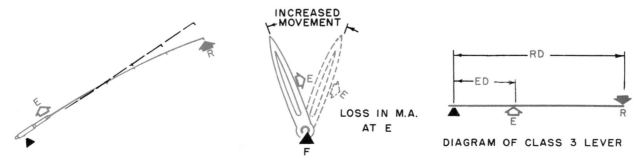

Fig. 11-4 Examples of third class levers

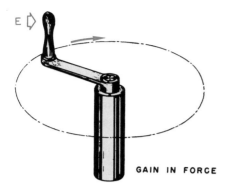

Fig. 11-5 Gaining mechanical advantage of force with the wheel axle

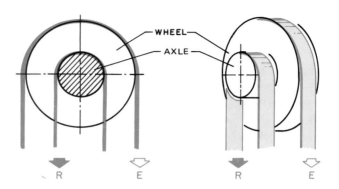

Metric

$$E \times ED = R \times RD$$

44.5 newtons ×

127 millimeters = R × 25.4 millimeters

5651.5 newton-millimeters = R

222.5 newtons = R

With a wheel and axle combination, a smaller force may be used to produce or control a larger force.

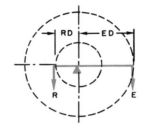

Fig. 11-6 Wheel and axle moments of force

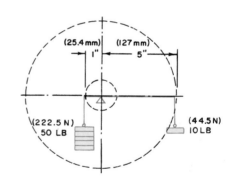

Fig. 11-7 Effort force of 10 pounds being applied to a 5-inch radius wheel

Inclined Plane

The inclined plane also allows force to be applied over a longer distance in order to handle heavy objects, figure 11-8. This principle as well as others was used by the Egyptians in the construction of the pyramids.

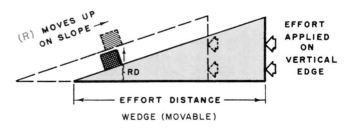

Fig. 11-8 The wedge is a variation of the inclined plane.

Stone blocks weighing several tons were rolled up a gently inclined plane and maneuvered into place. Upon completion of the structure, the earthen inclined planes were removed.

Inclined planes or ramps permit applying a lesser amount of force over a longer distance to ease the job at hand.

Example: a boy lifts his 100-pound (445-newton) body onto a 3 foot (0.9 meter) high loading dock with a single jump.

English

Work = Force × Distance

Work = 100 lb × 3 ft = 300 ft-lb

$$\text{Force} = \frac{\text{Work}}{\text{Distance}} = \frac{300 \text{ ft-lb}}{3 \text{ ft}} = 100 \text{ lb}$$

Metric

Work = Force × Distance

Work = 445 newtons × 0.9 meters = 400.5 joules

$$\text{Force} = \frac{\text{Work}}{\text{Distance}} = \frac{400.5 \text{ joules}}{0.9 \text{ meters}} = 445 \text{ newtons}$$

By walking up a 10 foot (3 meter) long ramp, less than the 100 pound (445 newton) force required in a single jump is applied, but it must be applied through a greater distance.

English

$$\text{Force} = \frac{\text{Work}}{\text{Distance}} = \frac{300 \text{ ft-lb}}{10 \text{ ft}} = 30 \text{ ft-lb when walking up ramp}$$

Metric

$$\text{Force} = \frac{\text{Work}}{\text{Distance}} = \frac{400.5 \text{ joules}}{3 \text{ meters}} = 133.5 \text{ newtons when walking up ramp}$$

In this example, the boy actually has a mechanical advantage of force when using the inclined plane. The mechanical advantage of force is 3 1/3.

English

$$\text{MA}_f = \frac{\text{R (Resistance Force)}}{\text{E (Effort Force)}}$$

$$\text{MA}_f = \frac{100 \text{ lb}}{30} = 3 \text{ 1/3}$$

Metric

$$\text{MA}_f = \frac{\text{R (Resistance Force)}}{\text{E (Effort Force)}}$$

$$\text{MA}_f = \frac{445 \text{ newtons}}{133.5} = 3 \text{ 1/3}$$

The mechanical advantage of distance can also be calculated.

English

$$\text{MA}_d = \frac{\text{RD (Resistance Distance)}}{\text{ED (Effort Distance)}}$$

$$\text{MA}_d = \frac{3 \text{ ft}}{10 \text{ ft}} = 0.3$$

Metric

$$\text{MA}_d = \frac{\text{RD (Resistance Distance)}}{\text{ED (Effort Distance)}}$$

$$\text{MA}_d = \frac{0.9 \text{ meters}}{3 \text{ meters}} = 0.3$$

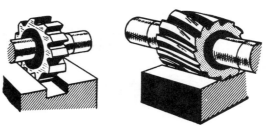

DETAIL OF INSIDE CUTTER

Fig. 11-9 Applications of the wedge as a separating device

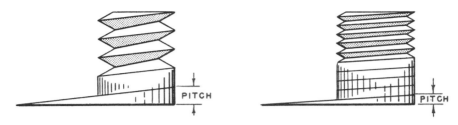

Fig. 11-10 The screw is an inclined plane in spiral form.

The mechanical advantage of speed is calculated with the same formula:

$$MA_s = \frac{RD}{ED}$$

Note that these mechanical advantage formulas apply to all simple machines.

Wedge.   The wedge is a variation of the inclined plane. It is driven under or into an object and applies its force through a longer distance. Many cutting tools are actually made up of a series of wedges which are called teeth. A conventional wedge and the wedge formed by a cutting tooth are the same; there is no difference in principle. Under magnification these wedge teeth get under and break away the material they are cutting, figure 11-9. Axes and chisels are also wedges. In addition, wedges make excellent holding devices. The simple doorstop is one example.

Screw.   The screw is also an inclined plane. However, it is one that has been put into a round form and appears as a continuous spiral, figure 11-10. Its most common application is in fasteners such as nuts, bolts, and wood screws. However, screw principles are used in jacks, worm gears, some pumps, and to convert rotary motion to linear motion. The mechanical advantage that a screw applies can be tremendous since the pitch of the

screw is usually quite small (RD) and the distance the effort is applied through (ED) is usually quite large by comparison. A jack screw illustrates this very well, figure 11-11.

Pulley

Pulleys in various arrangements create a mechanical advantage which can aid in the movement of heavy objects. In gaining the mechanical advantage, distance moved is sacrificed in the interest of applying less force. The number of ropes supporting the object determines the mechanical advantage. If four ropes support the weight, then the

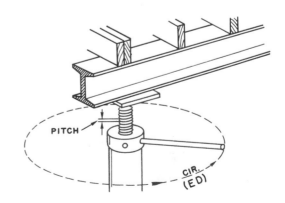

Fig. 11-11 The jack screw applies a tremendous mechanical advantage of force.

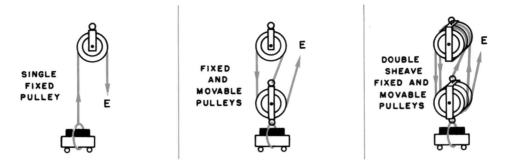

Fig. 11-12 Pulleys increase in mechanical advantage as the number of turns of rope (ED) increases.

rope is pulled four feet in order to move the object one foot. The force, disregarding friction, required is one-fourth the weight of the object, figure 11-12.

GEARS

The gear uses the same principles as simple machines. The gear is merely a wheel with special notches or teeth cut into its perimeter to provide positive contact with another gear. In a sense, two meshing gears are like two levers working against each other.

V-belts and pulleys, or chains and sprockets, have the same mechanical function as do gears. They are used when power is transmitted between shafts that are separated by a distance that make gears impractical. If construction permits close spacing of the shafts, then gear teeth can mesh against each other, a simpler arrangement.

Gears of different sizes change speed in proportion to the number of teeth on each gear, figure 11-13. A gear with 24 teeth will drive a gear with 12 teeth at double its speed.

$$MA_s = \frac{\text{Teeth in Driving Gear}}{\text{Teeth in Driven Gear}}$$

$$MA_s = \frac{24}{12} = 2$$

When speed changes, the torque also changes. Gearing up to increase speed reduces the torque produced. Gearing down to decrease speed increases torque. A gear with 24 teeth will drive a gear with 48 teeth at half its speed, but twice its force.

$$MA_f = \frac{\text{Teeth in Driven Gear}}{\text{Teeth in Driving Gear}}$$

$$MA_f = \frac{48}{24} = 2$$

The gearing in a motorcycle is an excellent example of this. At low speed it is geared down to gain the torque needed to get the motorcycle moving. At high speed, the motorcycle can be geared up. Less torque is needed, but more speed is required.

Gears also change the direction of rotation as the power is transmitted from one to another; from clockwise rotation to counterclockwise rotation and so forth, figure 11-14.

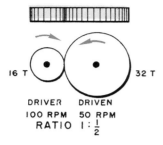

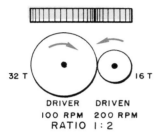

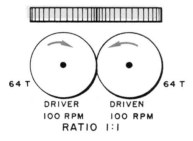

Fig. 11-13 Gears may be used to change speed and force.

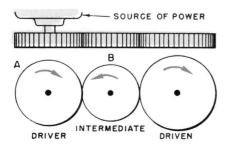

Fig. 11-14 Changing direction of rotation

THE POWER TRAIN

Many uses of small gasoline engines involve getting the power to the wheels as well as getting power to the machine. For example, a riding lawn mower engine must move the entire machine along as well as turn the blades for mowing grass. A wide variety of applications of levers, gears, pulleys, V-belts, shafts, chains, and sprockets are used to transmit the power of the engine's crankshaft at the correct speed and torque for the machine's requirements. Sometimes it is as simple as the direct drive of a rotary push-type mower. However, many uses are more complex and require a clutch, transmission, universal joint, and differential. These mechanisms are referred to as the power train.

- *Clutch* — connects and disconnects the power without stopping the engine.
- *Transmission* — provides for different speeds and directions (forward and reverse).
- *Universal Joint* — provides a flexible drive shaft.
- *Differential* — allows the wheels to rotate at different speeds for turning corners.

Clutch

The clutch engages or disengages the engine from the rest of the power train. When the engine is started and the machine is standing still, the clutch is disengaged. When the machine is moving, the clutch is engaged. Think of two pie plates, each on the end of a shaft, figure 11-15. One plate and shaft can rotate without the other plate and shaft rotating. However, if the rotating plate is moved into contact with and held against the nonrotating one, both will rotate.

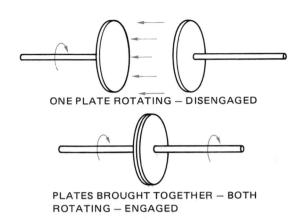

Fig. 11-15 When a nonrotating plate is held against a rotating plate, both plates will rotate.

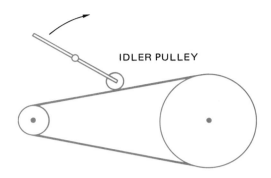

Fig. 11-16 Moving the idler pulley against the belt engages the drive.

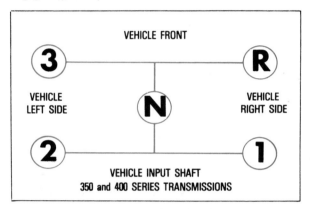

Fig. 11-17 Transmission shifting pattern

A very simple clutch arrangement is sometimes used on V-belts and pulleys. By having no belt tension the engine is disengaged. Tightening the belt tension with an idler pulley engages the machine, figure 11-16.

Transmission

Gears are arranged in a transmission in the proper sizes and combinations to satisfy the operating requirements of the machine that is being powered. Provisions are made to engage or shift the gears into various combinations which provide different output speeds and directions.

A typical transmission for small gasoline engines has three forward speeds and one reverse speed, figure 11-17. The Tecumseh/ Peerless model 350 shown in figure 11-18 is a good example of a typical small gasoline engine transmission. Power input is to the large bevel gear. A shifter fork mechanism shown in figure 11-19 slides the gears into mesh as required and power is transferred to the companion or output shaft. When the transmission is put into reverse, an idler gear

is brought into mesh. This "third" gear reverses the direction of rotation on the output shaft.

Universal Joint

Universal joints add flexibility to the mechanical transmission of power by connecting shafts that do not share the same axis, figure 11-20. The most common use for universal joints is between the transmission and differential. The universal joint compensates for misalignment, bouncing, and movement between the transmission and differential. The action of the universal joint is similar to the action of the human wrist. In addition to universal

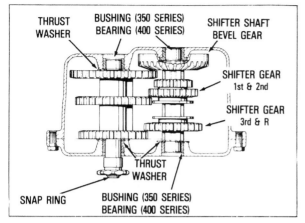

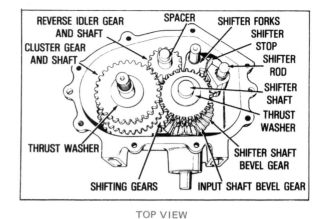

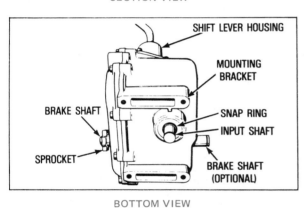

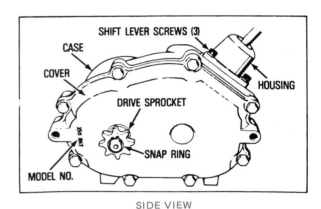

Fig. 11-18 A typical small gasoline engine transmission

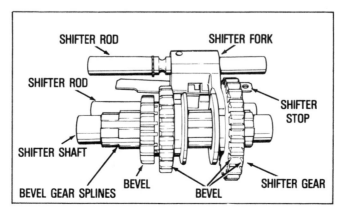

Fig. 11-19 Shifter fork, gears, and shaft

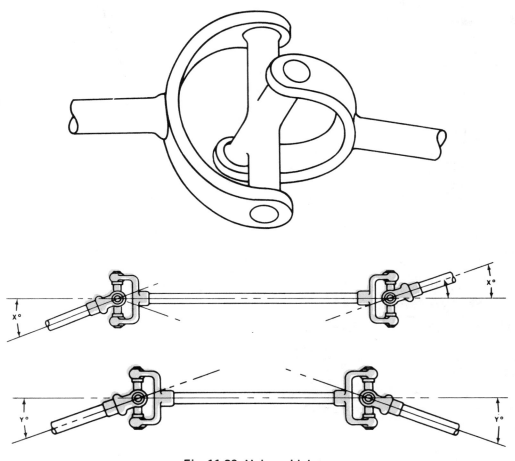

Fig. 11-20 Universal joint

joints, flexible couplings can be used to transmit power if there is a small amount of misalignment or motion between two units, figure 11-21.

Differential

The drive shaft and universal joints carry the power to the differential. The differential transmits the power to the axles and also enables the wheels to revolve at different speeds when a corner is turned, figure 11-22. The wheel on the inside of the turn needs to travel less distance than the wheel on the

outside of the turn, just as a column of marching soldiers do in turning a corner.

The parts of the differential are the case, drive pinion, ring gear, differential side bevel gears on the end of the axles, and pinion gears mounted in the differential case.

The drive pinion brings power into the differential and meshes with the ring gear. These two gears transfer the power from the drive shaft to the axles, which are at right angles to each other. The ring gear has about four times as many teeth as the drive pinion, the effect being a gear reduction ratio of

Fig. 11-21 Flexible couplings

4:1. To compensate for different rear wheel speeds when turning, the differential side bevel gears and pinion gears are added. These gears are mounted in the case. The ring gear is mounted on the case. When the vehicle travels straight ahead, the ring gear, case, and differential gears all move as a unit. However, when a corner is turned, the differential gears turn relative to each other; one faster, the other an equal amount slower. On slippery ice with one wheel spinning, the spinning wheel's differential side gear would be turning at twice the ring gear speed; the other differential side gear would be stationary.

Transaxles

Transaxles are a combination of the transmission, differential, and axles within one housing, figure 11-23. This results in a more compact and lighter weight unit. Transaxles are very common for riding lawnmowers, lawn tractors, and similar machines, figure 11-24. A transaxle unit with the top housing removed is shown in figure 11-25. The transaxle shown in figure 11-26 has five speeds forward and reverse. Power input is to the bevel gear. The shifter collar moves the

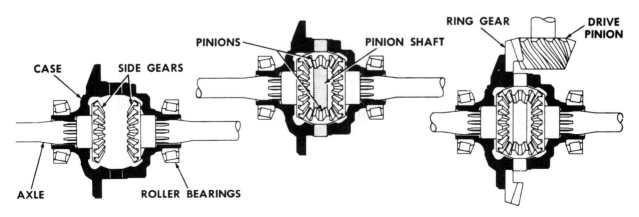

Fig. 11-22 The differential changes the direction of power

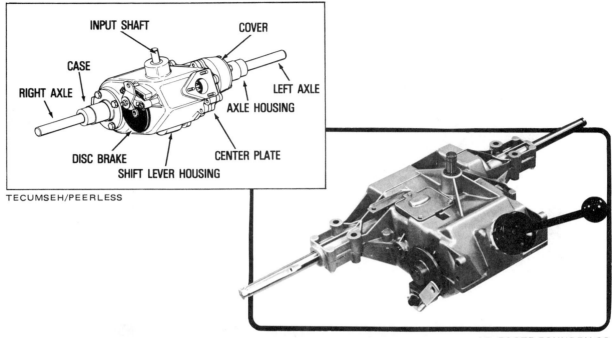

TECUMSEH/PEERLESS

J.B. FOOTE FOUNDRY CO.

Fig. 11-23 Transaxles

key into the shifter spur gear cluster to engage a particular gear. The remaining four shifter spur gears turn freely on the shifter and brake shaft. The spur gear at the end of the shifter and brake shaft turns the output gear; the output pinion engages the differential ring gear. The sprocket and chain drive are engaged for reverse.

Fig. 11-24 A transaxle allows the rear wheels to turn at different speeds as this tractor turns corners.

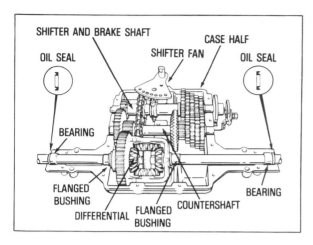

Fig. 11-25 Transaxle with top housing removed

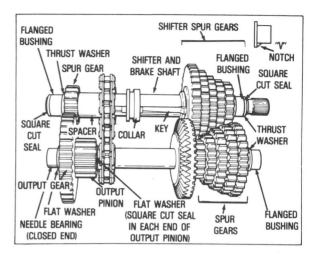

Fig. 11-26 Basic parts of a typical transmission portion of a transaxle

REVIEW QUESTIONS

1. Can direct or fixed transmission of power between the prime mover and the machine satisfy all machine applications? Explain.

2. What are the three basic power transmission systems?

3. List one example of each class of lever: first class, second class, third class.

4. Calculate the force necessary to move a 200-pound (890-newton) weight using a first class lever. The weight is 3 feet (0.9 meters) from the fulcrum. The lever's total length is 13 feet (4 meters).

5. Calculate the mechanical advantage of force found in carrying a 50-pound (222.5-newton) box up a 10-foot (3-meter) ramp that is 2 feet (0.6 meters) from ground level.

6. Compare a conventional wedge to the wedge formed by a cutting tooth.

7. What could be said about the mechanical advantage that is created by the use of the screw as a simple machine?

8. What determines the mechanical advantage of a pulley?

9. When are V-belts and pulleys or chains and sprockets used instead of gears?

10. Explain the relationship that exists between speed and torque.

11. What is the mechanical advantage of speed between a driven gear with 12 teeth and a driving gear with 96 teeth?

12. Explain the term *power train.*

13. What is the purpose of a clutch?

14. Explain why many engine applications need a transmission.

15. When might a universal joint be used?

16. Why do wheeled vehicles need a differential?

17. What is a transaxle?

CLASS DEMONSTRATION TOPICS

- Bring examples of levers to class. State the class of each lever and explain the mechanical advantage of force for each lever.

- Using a spring balance, weights of known value, and a yardstick for measuring, perform experiments using first, second, and third class levers.

- Bring in junked mechanical devices. Disassemble and analyze the working parts for example of simple machines.

- Using a spring balance, a coaster wagon, weights, and planks, demonstrate the mechanical advantage of an inclined plane, figure 11-27.

- Demonstrate the action of a universal joint using a universal joint from a socket wrench set.

- Set up various pulley systems to show the mechanical advantage of force that can be gained from pulleys, figure 11-28. Compare the distance the weight moves to the length of the rope pulled through the pulleys to determine the mechanical advantage of force. Use a spring scale to check the mechanical advantage of the pulley system.

- A bicycle is an excellent example of a power transmission system. Examine a five-speed bicycle. Be sure to point out the following:
  - The number of teeth on the front main sprocket.
  - The number of teeth on each of the gears on the rear sprocket.
  - The mechanical advantages of force for the various speeds of the bicycle.
  - How far the bicycle can travel in one revolution of the pedals, each speed.
  - The class of lever the caliper brakes represent.

- Disassemble and have the students inspect an automobile manual transmission. Perform experiments using the various gears.

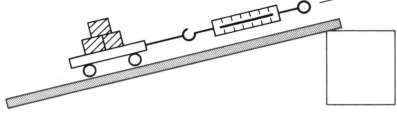

Fig. 11-27 Illustration of the mechanical advantage of an inclined plane

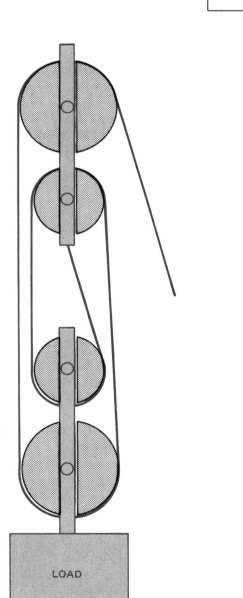

Fig. 11-28 Pulley systems are a mechanical advantage

LABORATORY EXPERIENCE 11-1
POWER TRAIN

The power train delivers power from the crankshaft of the engine to the work to be accomplished. Often requirements are complicated. A clutching device may be needed to disengage the engine from the machine without stopping the engine. A transmission provides for different speeds and may be used to change the direction of output shaft rotation. Universal joints may be used to transfer power through a distance when two shafts are not in alignment. Differentials compensate for different speeds of wheels during turning.

STUDENT ASSIGNMENT

You are to examine the power train from the engine's crankshaft to the final application of the power. Locate all of the basic parts. There are thousands of variations of power trains from different manufacturers and for different uses. If the instructor requires that you do actual disassembly work IT IS ESSENTIAL THAT YOU HAVE THE CORRECT MECHANIC'S REPAIR MANUAL AT HAND FOR THE MACHINE. To disassemble components of the power train without proper knowledge of disassembly and reassembly procedures can result in damage and frustration. Record your work in a work record box.

WORK RECORD BOX

PART	DISASSEMBLY (nuts, bolts, etc.)	OPERATION PERFORMED	TOOL USED

REVIEW QUESTIONS

1. Trace the path of the engine's power through the power train.

2. List the transmission principles that you observed — gears, levers, chains, pulleys, etc.

APPENDIX

COMMON SPECIFICATIONS FOR

1. **Spark plug gap: .030   All Models**
2. **Condenser capacity: .18 to .24 MFD. All Models**
3. **Contact point gap: .020   All Models**

	BASIC MODEL SERIES	IDLE SPEED	ARMATURE TWO LEG AIR GAP	THREE LEG AIR GAP	VALVE CLEARANCE INTAKE	EXHAUST	VALVE GUIDE REJECT GAGE	TORQUE SPECIFICATIONS FLYWHEEL NUT FT. LBS.	CYLINDER HEAD IN. LBS.	CONN. ROD IN. LBS.
A L U M I N U M	6B, 60000	1750	.006 .010	.012 .016	.005 .007	.009 .011	19122	55	140	100
	8B, 80000, 81000, 82000	1750	.006 .010	.012 .016	.005 .007	.009 .011	19122	55	140	100
	92000, 94000	1750	.006 .010		.005 .007	.009 .011	19122	55	140	100
	100000	1750	.010 .014	.012 .016	.005 .007	.009 .011	19122	60	140	100
	110000	1750	.006 .010		.005 .007	.009 .011	19122	55	140	100
	130000	1750	.010 .014		.005 .007	.009 .011	19122	60	140	100
	140000	1750	.010 .014	.016 .019	.005 .007	.009 .011	19151	65	165	165
	170000, 171700•	1750 **	.010 .014		.005 .007	.009 .011	19151	65	165	165
	190000, 191700•	1750 **	.010 .014		.005 .007	.009 .011	19151	65	165	165
	220000, 250000	1750 **	.010 .014		.005 .007	.009 .011	19151	65	165	190
C A S T I R O N	5, 6, N	1750		.012 .016	.007 .009	.014 .016	19122	55	140	100
	8	1750		.012 .016	.007 .009	.014 .016	19122	55	140	100
	9	1200			.007 .009	.014 .016	19151	60	140	140
	14	1200			.007 .009	.014 .016	19151	65	165	190
	19, 190000, 200000•	1200 **	.010 .014	.022 .026	.007 .009	.014 .016	19151	115	190	190
	23, 230000	1200 **	.010 .014	.022 .026	.007 .009	.017 .019	19151	145	190	190
	243000	1200 **	.010 .014		.007 .009	.017 .019	19151	145	190	190
	300000	1200 **	.010 .014		.007 .009	.017 .019	19151	145	190	190
	320000	1200 **	.010 .014		.007 .009	.017 .019	19151	145	190	190

COMMONLY USED TOOLS FOR SERVICING

19051	Spark Tester, all models	
19061	Carburetor jet screwdriver, all models	
19062	Carburetor jet screwdriver, all models	
19063	Valve spring compressor, all models	
† 19161	Starter clutch wrench, use with ½" drive torque wrench	
19203	Flywheel Puller, 170000 thru 320000 Aluminum Models & Cast Iron Models	

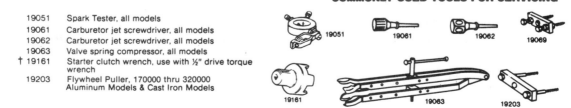

Fig. A-1 Engine specifications for all popular Briggs and Stratton engine models. A good service and repair manual contains specifications such as these for their engines.

ALL POPULAR ENGINE MODELS

4. Top governed speed: See Briggs & Stratton Service Bulletin No. 467 or Engine Replacement Data
5. Crankshaft End Play: .002-.008 All Models

CRANKSHAFT REJECT SIZE			MAIN BEARING REJECT GAGE	CYLINDER BORE STD. ▲	CARBURETOR TYPE	INITIAL CARBURETOR ADJUSTMENT ALL MODELS TURNS OPEN FROM SEAT	
MAG. JOURNAL	CRANKPIN	P.T.O. JOURNAL				NEEDLE VALVE	IDLE VALVE
.8726	.8697	.8726	19166	2.375* 2.374	Pulsa-Jet Vacu-Jet	1½	
.8726	.9963	.8726	19166	2.375 2.374	Two Piece Flo-Jet	1½	1
.8726	.9963	.8726	19166	2.5625 2.5615	One Piece Flo-Jet	2½	1½
.8726	.9963	.9976	19166 Mag. 19178 PTO	2.500 2.499	One Piece Flo-Jet (6, 7, 8, 10 & 11 H.P Vertical Crankshaft)	1½	1¼
.8726	.9963	.8726	19166	2.7812 2.7802	One Piece Flo-Jet (11 H.P. Horizontal Crankshaft)	1½	1
.8726	.9963	.9976	19166 Mag. 19178 PTO	2.5625 2.5615			
.9975	1.090	1.1790	19178	2.750 2.749			
.9975 1.1790•	1.090	1.1790	19178	3.000 2.999			
.9975 1.1790•	1.122	1.1790	19178	3.000 2.999			
1.3760	1.2470	1.3760		3.4375 3.4365			
.8726	.7433	.8726	19166	2.000 1.999			
.8726	.7433	.8726	19166	2.250 2.249			
.9832	.8726	.9832	19117	2.250 2.249			
1.1790	.9964	1.1790	19117	2.625 2.624			
1.1800	.9964 1.1219•	1.1790	19117	3.000 2.999			
1.3769	1.1844	1.3759	19117	3.000 2.999			
Ball	1.3094	Ball	Ball	3.0625 3.0615			
Ball	1.3094	Ball	Ball	3.4375 3.4365			
Ball	1.3094	Ball	Ball	3.5625 3.5615			

CYLINDER RESIZING

▲ Resize if .003 or more wear or .0015 out of round on C.I. cylinder engines, .0025 out of round on aluminum alloy engines.

Resize to .010, .020 or .030 over Standard.

*Model 6B series and early Models 60000 and 61000 series engines have cylinder bore of 2.3125 – 2.3115.

RING GAP REJECT SIZES

MODEL	COMP. RINGS	OIL RING
Alum. Cylinder Models	.035"	.045"
C.I. Cylinder Models	.030"	.035"

**GOVERNED IDLE

For Adjustment Procedures, see Service & Repair Instructions 270962, Section 5, for Single Cylinder Models and Repair Instructions MS-7000 or 271172, Section 5, for Twin Cylinder Models.

■ With Valve Springs Installed.

• Synchro-Balance.

BRIGGS & STRATTON ENGINES

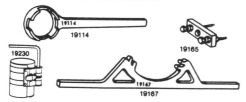

Flywheel puller, all models thru 130000	19069
Piston ring compressor, all models	19230
Starter clutch wrench, all rewind starter models	19114
Flywheel Puller, 140000, 170000, 190000 & 250000 Models	19165
Flywheel holder, all models 6B thru 130000	19167

See Section 13 for Complete List of Tools

GENERAL INFORMATION
Check-up

CHECK - UP

Most complaints concerning engine operation can be classified as one or a combination of the following:

1. Will not start

2. Hard starting

3. Kicks back when starting

4. Lack of power

5. Vibration

6. Erratic operation

7. Overheating

8. High oil consumption

When the cause of malfunction is not readily apparent, perform a check of the Compression, Ignition and Carburetion Systems. This check-up, performed in a systematic manner, can usually be done in a matter of minutes. It is the quickest and surest method of determining the cause of failure. This check-up will point up possible cause of future failures, which can be corrected at the time. The basic check-up procedure is the same for all engine models, while any variation, by model will be shown under the subject heading.

NOTE: What appears to be an engine malfunction may be a fault of the powered equipment rather than the engine. If equipment is suspect, see Equipment, affecting engine operation.

Check Compression

Spin flywheel in reverse rotation (counterclockwise) to obtain accurate compression check. The flywheel should rebound sharply, indicating satisfactory compression.

If compression is poor, look for —

1. Loose spark plug

2. Loose cylinder head bolts

3. Blown head gasket

4. Burnt valves and/or seats

5. Insufficient tappet clearance

6. Warped cylinder head

7. Warped valve stems

8. Worn bore and/or rings

9. Broken connecting rod

Check Ignition

Remove the spark plug. Spin the flywheel rapidly with one end of the ignition cable clipped to the 19051 tester and with the other end of the tester grounded on the cylinder head. If spark jumps the .166" tester gap, you may assume the ignition system is functioning satisfactorily. Try a new spark plug.

If spark does not occur look for —
1. Incorrect armature air gap
2. Worn bearings and/or shaft on flywheel side
3. Sheared flywheel key
4. Incorrect breaker point gap
5. Dirty or burned breaker points
6. Breaker plunger stuck or worn
7. Shorted ground wire (when so equipped)
8. Shorted stop switch (when so equipped)
9. Condenser failure
10. Armature failure
11. Improperly operating interlock system

NOTE: If engine runs but misses during operation, a quick check to determine if ignition is or is not at fault can be made by inserting the 19051 tester between the ignition cable and the spark plug. A spark miss will be readily apparent. While conducting this test on Magna-Matic equipped engines, Models 9, 14, 19 and 23, set the tester gap at .060".

Fig. A-2 General check-up procedures (troubleshooting) suggested for Briggs and Stratton engines.

GENERAL INFORMATION
Check-up

Check Carburetion

Before making a carburetion check, be sure the fuel tank has an ample supply of fresh, clean gasoline. On gravity feed (Flo-Jet) models, see that the shut-off valve is open and fuel flows freely through the fuel line. On all models, inspect and adjust the needle valves. Check to see that the choke closes completely. If engine will not start, remove and inspect the spark plug. If plug is wet, look for —

1. Overchoking
2. Excessively rich fuel mixture
3. Water in fuel
4. Inlet valve stuck open (Flo-Jet carburetor)

If plug is dry, look for —

1. Leaking carburetor mounting gaskets
2. Gummy or dirty screen or check valve (Pulsa-Jet and Vacu-Jet carburetors)
3. Inlet valve stuck shut (Flo-Jet carburetors)
4. Inoperative pump (Pulsa-Jet carburetors)

A simple check to determine if the fuel is getting to the combustion chamber through the carburetor is to remove the spark plug and pour a small quantity of gasoline through the spark plug hole. Replace the plug. If the engine fires a few times and then quits, look for the same condition as for a dry plug.

Equipment - Effecting Engine Operation

Frequently, what appears to be a problem with engine operations, such as hard starting, vibration, etc., may be the fault of the equipment powered rather than the engine itself. Since many varied types of equipment are powered by Briggs and Stratton engines, it is not possible to list all of the various conditions that may exist. Listed are the most common effects of equipment problems, and what to look for as the most common cause.

Hard Starting, Kickback, or Will Not Start

1. Loose blade — Blade must be tight to shaft or adaptor.
2. Loose belt — a loose belt like a loose blade can cause a back-lash effect, which will counteract engine cranking effort.
3. Starting under load — See if the unit is disengaged when engine is started; or if engaged, does not have a heavy starting load.
4. Check remote Choke-A-Matic control assembly for proper adjustment.
5. Check interlock system for shorted wires, loose or corroded connections, or defective modules or switches.

Vibration

1. Cutter blade bent or out of balance — Remove and balance
2. Crankshaft bent — Replace
3. Worn blade coupling — Replace if coupling allows blade to shift, causing unbalance.
4. Mounting bolts loose — Tighten
5. Mounting deck or plate cracked — Repair or replace.

Power Loss

1. Bind or drag in unit — If possible, disengage engine and operate unit manually to feel for any binding action.
2. Grass cuttings build-up under deck.
3. No lubrication in transmission or gear box.
4. Excessive drive belt tension may cause seizure.

Noise

1. Cutter blade coupling or pulley — an oversize or worn coupling can result in knocking, usually under acceleration. Check for fit, or tightness.
2. No lubricant in transmission or gear box.

Specifications/model K91

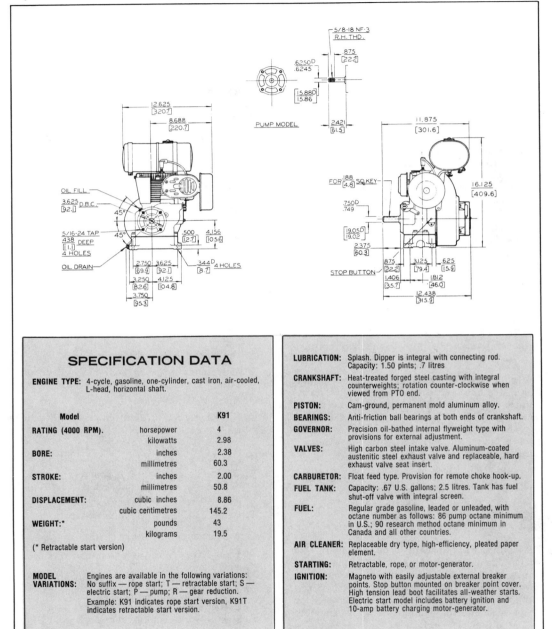

SPECIFICATION DATA

ENGINE TYPE: 4-cycle, gasoline, one-cylinder, cast iron, air-cooled, L-head, horizontal shaft.

Model		K91
RATING (4000 RPM).	horsepower	4
	kilowatts	2.98
BORE:	inches	2.38
	millimetres	60.3
STROKE:	inches	2.00
	millimetres	50.8
DISPLACEMENT:	cubic inches	8.86
	cubic centimetres	145.2
WEIGHT:*	pounds	43
	kilograms	19.5

(* Retractable start version)

MODEL VARIATIONS: Engines are available in the following variations: No suffix — rope start; T — retractable start; S — electric start; P — pump; R — gear reduction. Example: K91 indicates rope start version, K91T indicates retractable start version.

LUBRICATION: Splash. Dipper is integral with connecting rod. Capacity: 1.50 pints; .7 litres

CRANKSHAFT: Heat-treated forged steel casting with integral counterweights; rotation counter-clockwise when viewed from PTO end.

PISTON: Cam-ground, permanent mold aluminum alloy.

BEARINGS: Anti-friction ball bearings at both ends of crankshaft.

GOVERNOR: Precision oil-bathed internal flyweight type with provisions for external adjustment.

VALVES: High carbon steel intake valve. Aluminum-coated austenitic steel exhaust valve and replaceable, hard exhaust valve seat insert.

CARBURETOR: Float feed type. Provision for remote choke hook-up.

FUEL TANK: Capacity: .67 U.S. gallons; 2.5 litres. Tank has fuel shut-off valve with integral screen.

FUEL: Regular grade gasoline, leaded or unleaded, with octane number as follows: 86 pump octane minimum in U.S.; 90 research method octane minimum in Canada and all other countries.

AIR CLEANER: Replaceable dry type, high-efficiency, pleated paper element.

STARTING: Retractable, rope, or motor-generator.

IGNITION: Magneto with easily adjustable external breaker points. Stop button mounted on breaker point cover. High tension lead boot facilitates all-weather starts. Electric start model includes battery ignition and 10-amp battery charging motor-generator.

KOHLER

Fig. A-3 Specification sheet for Kohler Model K91. Specification sheets like these help the engineer, designer, and buyer select the type of engine suitable for a particular application or job.

DISASSEMBLY and REASSEMBLY PROCEDURES

With the disassembly operations, instructions on reassembly are also given, as often it will not be necessary to disassemble the entire engine. If it is desired to disassemble the entire engine, the reassembly instructions can be looked up under the headings of the various parts.

Tighten the capscrews of the cylinder head, gear cover, connecting rod, flywheel and spark plug to the specified torque readings indicated in the related paragraphs of reassembly. Use caution in tightening all other screws and nuts so as not to strip threads or break flanges.

RECOIL STARTER

1. Remove four mounting screws and take off recoil starter assembly.

2. *In reassembly;* check condition of rope for possible replacement. Repair per the *Repair Instructions* beginning on *Page 20*.

FLYWHEEL SHROUD

1. Disconnect coil primary wire from stop button wire.

2. Take off flywheel shroud from cylinder-case and cylinder head.

3. Remove baffle from cylinder block.

Fig. A-4 Disassembly and reassembly procedures for a Wisconsin ROBIN engine Model W1-145. This is an excellent example of the type of detailed information that a good repair manual will provide the mechanic.

FUEL TANK

1. Disconnect fuel line at carburetor. *In reassembly;* it is advisable to use a new fuel line.

2. Remove fuel tank from cylinder head.

3. Remove cover from cylinder head.

AIR CLEANER

1. Remove air cleaner cover, element and retainer.

2. Disconnect breather tube from inspection cover.

3. Take out two lock nuts and remove air cleaner body and gasket from carburetor studs.

4. *In reassembly;* wash element per *Air Cleaner Instructions* on *Page 4*. Mount new gasket and assemble back cover, elements and cover per *Fig. 3*. Connect breather line.

GOVERNOR LEVER and CARBURETOR

1. Disconnect governor spring from lever to speed control assembly.

2. Remove governor lever from shaft, and disconnect rod and spring from lever to carburetor.

3. Remove carburetor, insulating plate and gaskets from studs in cylinder-case.

4. If necessary, the speed control assembly can be removed from cylinder head by taking off flange bolt.

5. *In reassembly;* see 'Governor Adjustment', Page 6.

MUFFLER

Take off the two hex nuts and remove muffler and gasket from cylinder-case.

STARTING PULLEY and FLYWHEEL (magneto)

1. Place a *19mm* socket wrench on to the flywheel nut and give the wrench a sharp blow with a soft hammer. Remove nut, spring washer and pulley.

2. As illustrated in *Fig. 10,* attach puller to flywheel — turn center bolt clockwise until flywheel becomes loose enough to be removed.

   If puller is not available: Screw flywheel nut flush with end of crankshaft to protect shaft threads from being

damaged. Place the end of a large screwdriver between the crankcase and flywheel in line with the keyway. Then, strike the end of the flywheel nut with a babbitt hammer and at the same time wedge outward with the screwdriver.

3. The ignition coil along with attached high tension cable can then be removed by taking out two mounting screws.

4. Take off point cover and remove contact breaker and condenser.

5. *In reassembly;* refer to *'Breaker Point Adjustment'* and *'Timing'* instructions.

6. Securely tighten flywheel nut after timing is finalized, but first be sure woodruff key is in position on shaft. *Do not* drive flywheel on to taper of crankshaft and *do not* overtighten flywheel nut. Simply turn nut until lockwasher collapses. Then, tighten by placing wrench on nut and giving handle of wrench 1 or 2 sharp blows with a soft hammer. If torque wrench is available, tighten *44* to *47 ft. lbs.* (6 to 6.5 kgm).

CYLINDER HEAD and SPARK PLUG

1. Remove spark plug from cylinder head.

2. Loosen mounting screws and remove cylinder head along with gasket.

3. Clean carbon from combustion chamber and dirt from between the cooling fins. Check the cylinder head mounting face for distortion. If *warpage* is evident, replace head.

4. *In reassembly;* use new cylinder head gasket and spark plug. *Torque* head screws *14* to *17 ft. lbs.* (1.9 to 2.3 kgm).

   Leave spark plug out temporarily, for ease in turning engine over for remainder of assembly and for timing adjustments. When mounting spark plug, tighten *8.7* to *10.8 ft. lbs.* (1.2 to 1.5 kgm).

INTAKE and EXHAUST VALVES, Fig's. 11, 12, 13

1. Remove the valve inspection cover, breather plate and gaskets from cylinder-case.

2. Position valve spring retainer so that *notch* on outer diameter is toward the outside of the valve compartment.

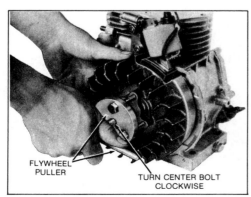

FLYWHEEL PULLER

TURN CENTER BOLT CLOCKWISE

Fig. 10

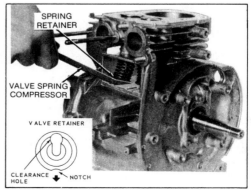

SPRING RETAINER

VALVE SPRING COMPRESSOR

VALVE RETAINER

CLEARANCE HOLE — NOTCH

Fig. 11

Place **compressor tool** EY2079500307 under spring retainer and compress valve spring. At the same time, draw retainer outward toward notch so that **clearance hole** in retainer allows for valve to be pulled out from top of cylinder block.

Caution: **Do not damage** gasket surface of tappet chamber with the compressor tool.

3. Clean carbon and gum deposits from the valves, seats, ports and guides.

4. **In reassembly;** replace valves that are badly burned, pitted or warped. With retainer notch facing outward, place retainer under valve spring with clearance hole over valve stem. While compressing valve spring with compressor tool, force retainer inward so that it locks in place on valve stem. *See Fig. 11.*

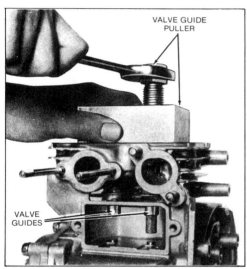

Fig. 12

5. **Valve guides** should be replaced when valve stem clearance becomes excessive. Use valve guide puller tool **EY2279500107** as illustrated in *Fig. 12.* Draw valve guides out and press new guides in, using the same puller tool. Refer to *Fig. 13* for clearance specifications and proper assembly.

6. After replacing valves and valve guides, lap valves in place until a uniform ring will show entirely around the face of the valve. Clean valves, and wash block thoroughly with a hot solution of soap and water. Wipe cylinder walls with clean lint free rags and light engine oil. **Do not assemble valve springs** until tappet clearance has been checked. *See 'Tappet Adjustment'.*

TAPPET ADJUSTMENT, Fig. 14

With tappet in its lowest position, hold valve down and insert feeler gauge between valve and tappet stem. The clearance for both intake and exhaust, with engine *cold:* **0.003 to 0.005 inch** (0.08 to 0.12mm).

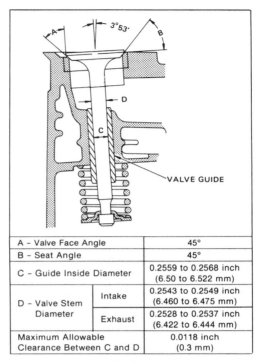

A – Valve Face Angle		45°
B – Seat Angle		45°
C – Guide Inside Diameter		0.2559 to 0.2568 inch (6.50 to 6.522 mm)
D – Valve Stem Diameter	Intake	0.2543 to 0.2549 inch (6.460 to 6.475 mm)
	Exhaust	0.2528 to 0.2537 inch (6.422 to 6.444 mm)
Maximum Allowable Clearance Between C and D		0.0118 inch (0.3 mm)

Fig. 13

If the clearance is less than it should be, grind the end of valve stem a very little at a time and remeasure. Stems must be ground square and flat.

If clearance is too large, replace the valve and, or tappet.

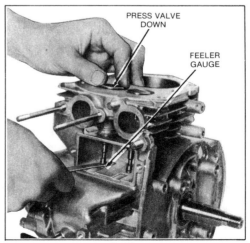

Fig. 14

After obtaining correct clearance, assemble valve springs and retainers. Check operation of valves by turning crankshaft over — remeasure tappet clearance.

GEAR COVER, Fig's. 15, 16 and 17

Place a rag under engine to absorb excess oil, and remove gear cover mounting screws. Use a soft hammer and evenly tap around outer surface of gear cover until it breaks free from crankcase face. Carefully remove cover so as not to damage oil seal. Dowel pins will remain in crankcase flange.

Inspect adjusting shim, oil seal, governor cross shaft, gear and thrust sleeve. If any of these parts are damaged or worn — they should be replaced.

Governor Gear-Flyweight Assembly; can be removed by straddling the handles of a pliers under the gear hub and across the gear cover face. Then, press down on the jaw end of the pliers and gear-thrust sleeve will loosen from shaft. *In Reassembly;* place spacing washer on shaft. Insert sleeve between flyweights, mount complete assembly to shaft and press down until unit *snaps* in place. *NOTE: Thrust sleeve cannot be assembled or removed when gear assembly is mounted to shaft.*

Crankshaft End Play is regulated by a selected thickness of adjusting shim. The *end play* should be *0 to 0.008 inch* (0 to 0.2mm) *engine cold.*

The *adjusting shim* is located between the crankshaft gear and main bearing, and replacement is seldom necessary unless the crankshaft or gear cover are replaced.

To determine what thickness *adjusting shim* to use for regulating the end play, refer to *Fig. 15* and the following instructions:

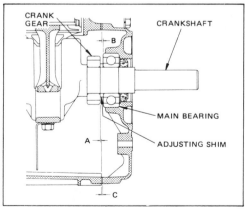

Fig. 15

1. With the gear cover removed, tap end of crankshaft slightly to insure that shaft is shouldered against main bearing at flywheel end.

2. Use a Depth Micrometer and measure the distance between the machined surface of crankcase face and end of crank gear (dim. *A*).

3. Measure the distance between the machined face of gear cover and outer face of main bearing (dim. *B*).

4. The compressed thickness of gear cover gasket is *0.0086 inch* (0.22mm) dim. *C*.

5. From the following chart, select an *adjusting shim* that is *0 to 0.008 inch* (0 to 0.2mm) *less* that the total length, *A + B + C*.

Adjusting Shim Thickness	Part Number
.0215 to .0257 inch (.546 to .654 mm)	EY0230250110
.0287 to .0343 inch (.728 to .872 mm)	EY0230250120
.0362 to .0425 inch (.920 to 1.080 mm)	EY0230250130

In reassembly of gear cover, apply oil to bearing surfaces, gear train and tappets. Also lubricate lips of oil seal and add a light film of oil on gear cover face to hold gasket in place.

Mount adjusting shim to crankshaft at crank gear face.

When assembling gear cover, as illustrated in *Fig. 16*, be sure that governor gear-flyweight assembly is in place, and that thrust sleeve aligns with governor cross shaft. If available, use an oil seal sleeve mounted on to the crankshaft, to prevent damage to the oil seal lips.

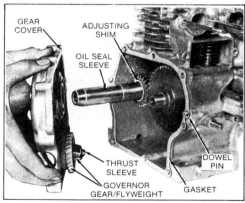

Fig. 16

Caution: Be sure timing marks on crankshaft and camshaft gear, Fig. 18, remain correctly mated when end of camshaft is fitted into bearing hole of gear cover, and that governor gear engages camshaft and gear.

Tap gear cover in place with a soft hammer, remove oil seal sleeve and tighten gear cover capscrews to *5.8 to 7.2 ft. lbs. torque* (0.8 to 1.0 Kgm).

Check End Play after gear cover and flywheel are in place. If the proper thickness of adjusting shim was selected as explained in *'Crankshaft End Play'* paragraphs and *Fig. 15,* then a satisfactory end play of *0 to 0.008 inch* (0 to 0.2mm) should have been obtained. However, on instal-

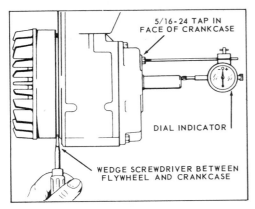

5/16-24 TAP IN
FACE OF CRANKCASE

DIAL INDICATOR

WEDGE SCREWDRIVER BETWEEN
FLYWHEEL AND CRANKCASE

Fig. 17

lations where end play is critical, and to confirm the actual amount of end play, adhere to the following instructions and *Fig. 17.*

1. Tap end of crankshaft with a soft hammer so that crankshaft will shoulder against main bearing at flywheel end.

2. Tap crankshaft in the opposite direction (from flywheel end) to seat adjusting shim against main bearing at take-off end.

3. Attach a Dial Indicator to one of the 5/16-24 tapped holes on the face of the crankcase, and set the dial at 0.

4. Wedge a screwdriver between the flywheel and crankcase. The movement of the flywheel away from the engine block will register as end play on the Indicator dial.

If *end play* is *more* than the limits of *0* to *0.008 inch* (0 to 0.2mm) use the reading from the Indicator dial as a guide in selecting what thickness of adjusting shim to use. If there is *no end play*, revert to *Gear Cover* paragraphs on Page 12 and follow procedures for adjusting *crankshaft end play* and determine exactly what adjusting shim thickness is required.

CAMSHAFT, TAPPETS and TIMING MARKS, Fig. 18

1. To prevent tappets from falling out and becoming damaged when camshaft is removed, turn crankcase over on its side as shown in *Fig. 18.* Push tappets inward to clear cam lobes, and remove camshaft.

2. Withdraw tappets and mark them for identification with the hole that they were removed from.

3. *In reassembly,* put tappets back in their corresponding guide hole. This will eliminate unnecessary valve stem grinding for obtaining correct tappet clearance.

4. Lubricate bearing surfaces of camshaft, push valve tappets to their uppermost position and mount camshaft assembly to crankcase.

5. *Timing marks* on camshaft gear and crankshaft gear must match up. Mount camshaft so that the *marked tooth on crankshaft gear* is between the *two marked teeth* of the *camshaft gear, Fig. 18.* If valve timing is off, engine will not function properly or may not run at all.

CONNECTING ROD and PISTON, Fig's. 19 thru 24

1. Straighten out the bent tabs of lock plate and remove bolts from connecting rod.

2. Take off lock plate, oil dipper and connecting rod cap.

3. Scrape off all carbon deposits that might interfere with removal of piston from upper end of cylinder.

4. Turn crankshaft until piston is at top, then push connecting rod and piston assembly upward and out thru top of cylinder.

5. Remove piston from connecting rod by taking out one of the snap rings and then removing the piston pin. *A new snap ring should be used in reassembly.*

6. *Reassembly*

   a. *Piston Rings*

   As illustrated in *Fig. 19,* use a Ring Expander tool to prevent ring from becoming distorted or broken when installing onto piston.

   If an expander tool is not available, install rings by placing the open end of the ring on first land of

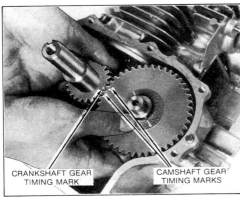

CRANKSHAFT GEAR
TIMING MARK

CAMSHAFT GEAR
TIMING MARKS

Fig. 18

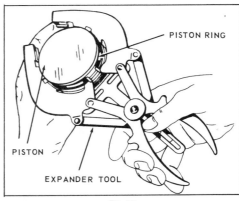

PISTON RING

PISTON

EXPANDER TOOL

Fig. 19

piston. Spread ring only far enough to slip over piston and into correct groove, being careful not to distort ring.

With or without expander tool, assemble bottom ring first and work upward, installing top ring last.

Mount scraper ring with scraper edge down, otherwise oil pumping and excessive oil consumption will result. Refer to *Fig. 20* for correct placement of rings.

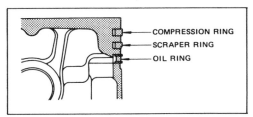

COMPRESSION RING
SCRAPER RING
OIL RING

Fig. 20

b. *Piston, Cylinder and Connecting Rod*

Measure the diameter of the piston in the center of the thrust faces at the bottom of the piston skirt, as illustrated in *Fig. 21*.

Measure the cylinder bore and inspect for out-of-round and taper. If cylinder is scored or worn more than *.005"* (0.125mm) over standard size, it should be rebored and fitted with an oversize piston and rings.

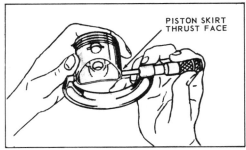

PISTON SKIRT
THRUST FACE

Fig. 21

Refer to chart, *Fig. 22,* for clearance between piston and cylinder. Size, clearance and wear limits are given in more detail on *Page 17.*

When installing the piston in the cylinder, oil the piston, rings, wrist pin, rod bearings and cylinder wall before assembly. Stagger the piston ring gaps 90° apart around the piston. Use a piston ring compressor as illustrated in *Fig. 23.*

Turn crankshaft to bottom of stroke and tap piston down until rod contacts crank pin. Mount connecting rod cap so that the *cast rib* between face of rod and bolt boss matches up with the cast rib on the connecting rod. *See Fig. 24.* Assemble *oil dipper* to cap. The dipper should be *toward the flywheel* end of the connecting rod cap. If engine is operated on a *tilt toward the take-off end*, reverse dipper mounting.

		Inches	Millimeters
Piston to Cylinder at Piston Skirt Thrust Face		0.0008 0.0023	0.020 0.059
Connecting Rod to Crank Pin	Dia.	0.0015 0.0025	0.037 0.063
	Side	0.0039 0.0118	0.1 0.3
Piston Pin to Connecting Rod		0.0004 0.0012	0.010 0.029
D – Crankshaft Pin Diameter		0.9429 0.9434	23.950 23.963
W – Crankshaft Pin Width		0.9055 0.9094	23.0 23.1
Piston Ring Gap		0.008 0.016	0.2 0.4
Piston Rings – Side Clearance In Grooves	Top	0.0035 0.0053	0.090 0.135
	2nd	0.0024 0.0041	0.060 0.105
	Oil	0.0004 0.0025	0.010 0.065
Piston Pin to Piston		0.00035 tight to 0.00039 loose	0.009 0.010

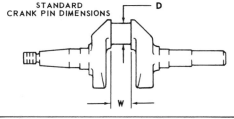

STANDARD CRANK PIN DIMENSIONS

D

W

Fig. 22 PISTON, RING AND ROD CLEARANCE CHART

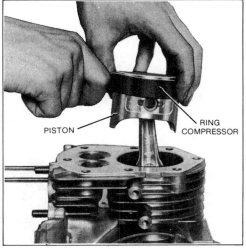

PISTON

RING COMPRESSOR

Fig. 23

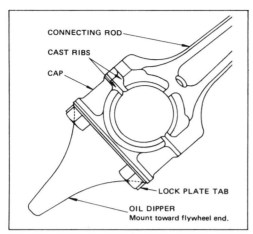

CONNECTING ROD

CAST RIBS

CAP

LOCK PLATE TAB

OIL DIPPER
Mount toward flywheel end.

Fig. 24

Install a new rod bolt lock plate. Mount connecting rod bolts and tighten to the following torque specifications — *6.5* to *8.3 ft. lbs.* (0.90 to 1.15 Kgm).

Check for free movement of connecting rod by turning crankshaft over slowly. If satisfactory, bend tabs on lock plate against hex head flat of connecting rod bolts.

CRANKSHAFT, Fig. 25

1. Remove flywheel woodruff key.

2. Pull crankshaft out from open end of crankcase and take care not to damage the oil seal. If necessary, loosen shaft by tapping lightly at flywheel end with a soft hammer.

3. *In reassembly,* inspect crankcase oil seal and main bearing for possible replacement. Mount crankshaft with extreme care so as not to damage lips of oil seal. Use an oil seal sleeve if available.

4. *End Play* is regulated by means of an *adjusting shim* at the gear end of the crankshaft. This should be set immediately before mounting gear cover as explained in *'Crankshaft End Play'* paragraphs on *Page 12*.

CRANKSHAFT

Fig. 25

ENGINE REASSEMBLY

Rebuild the engine in reverse order, and by following the reassembly instructions which are included with the disassembly procedures.

After complete reassembly, turn the engine over by means of the starting pulley to check for any abnormal conditions and loose fitting parts. Then, review and proceed with the following:

1. Crankshaft end play — *Page 12*.

2. Check ignition spark — *Page 4*.

3. Timing — *Page 5*.

4. Governor adjustment — *Page 6*.

5. Carburetor adjustments — *Page 20*.

6. Fill crankcase with correct grade of oil — *Page 4*.

7. Fill fuel tank with a well known brand of *'Regular Grade'* gasoline.

8. Test engine per the following instructions.

TESTING REBUILT ENGINE

An engine that has been completely overhauled by having the cylinder rebored and fitted with a new piston, rings, valves and connecting rod should be thoroughly *'RUN-IN'* before being put back into service. Good bearing surfaces and running clearances between the various parts can only be established by operating the engine under reduced speed and loads for a short period of time.

Load		Speed	Time
No Load		2500 r.p.m.	10 minutes
No Load		3000 r.p.m.	10 minutes
No Load		3600 r.p.m.	10 minutes
1.35 hp.	1.01 kw.	3600 r.p.m.	30 minutes
2.7 hp.	2.01 kw.	3600 r.p.m.	60 minutes

While engine is being tested — check for oil leaks. Make final carburetor adjustments and regulate the engine operating speed.

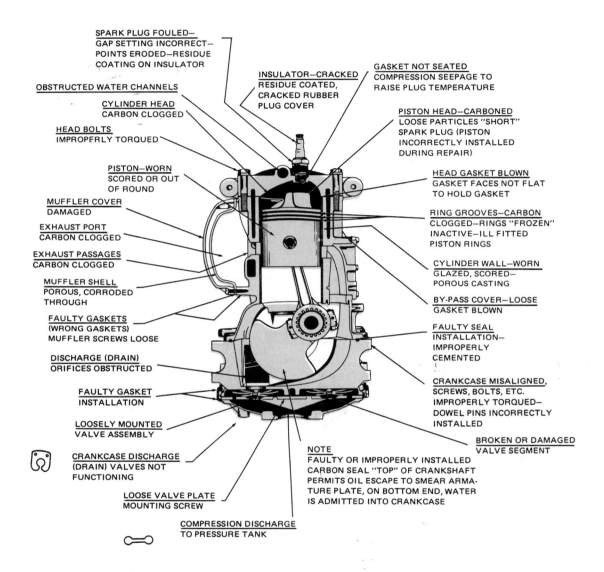

SPARK PLUG FOULED—
GAP SETTING INCORRECT—
POINTS ERODED—RESIDUE
COATING ON INSULATOR

INSULATOR—CRACKED
RESIDUE COATED,
CRACKED RUBBER
PLUG COVER

GASKET NOT SEATED
COMPRESSION SEEPAGE TO
RAISE PLUG TEMPERATURE

OBSTRUCTED WATER CHANNELS

CYLINDER HEAD
CARBON CLOGGED

PISTON HEAD—CARBONED
LOOSE PARTICLES "SHORT"
SPARK PLUG (PISTON
INCORRECTLY INSTALLED
DURING REPAIR)

HEAD BOLTS
IMPROPERLY TORQUED

PISTON—WORN
SCORED OR OUT
OF ROUND

HEAD GASKET BLOWN
GASKET FACES NOT FLAT
TO HOLD GASKET

MUFFLER COVER
DAMAGED

RING GROOVES—CARBON
CLOGGED—RINGS "FROZEN"
INACTIVE—ILL FITTED
PISTON RINGS

EXHAUST PORT
CARBON CLOGGED

EXHAUST PASSAGES
CARBON CLOGGED

CYLINDER WALL—WORN
GLAZED, SCORED—
POROUS CASTING

MUFFLER SHELL
POROUS, CORRODED
THROUGH

BY-PASS COVER—LOOSE
GASKET BLOWN

FAULTY GASKETS
(WRONG GASKETS)
MUFFLER SCREWS LOOSE

FAULTY SEAL
INSTALLATION—
IMPROPERLY
CEMENTED

DISCHARGE (DRAIN)
ORIFICES OBSTRUCTED

CRANKCASE MISALIGNED,
SCREWS, BOLTS, ETC.
IMPROPERLY TORQUED—
DOWEL PINS INCORRECTLY
INSTALLED

FAULTY GASKET
INSTALLATION

LOOSELY MOUNTED
VALVE ASSEMBLY

BROKEN OR DAMAGED
VALVE SEGMENT

CRANKCASE DISCHARGE
(DRAIN) VALVES NOT
FUNCTIONING

NOTE
FAULTY OR IMPROPERLY INSTALLED
CARBON SEAL "TOP" OF CRANKSHAFT
PERMITS OIL ESCAPE TO SMEAR ARMA-
TURE PLATE, ON BOTTOM END, WATER
IS ADMITTED INTO CRANKCASE

LOOSE VALVE PLATE
MOUNTING SCREW

COMPRESSION DISCHARGE
TO PRESSURE TANK

Fig. A-5 Power head diagnosis chart

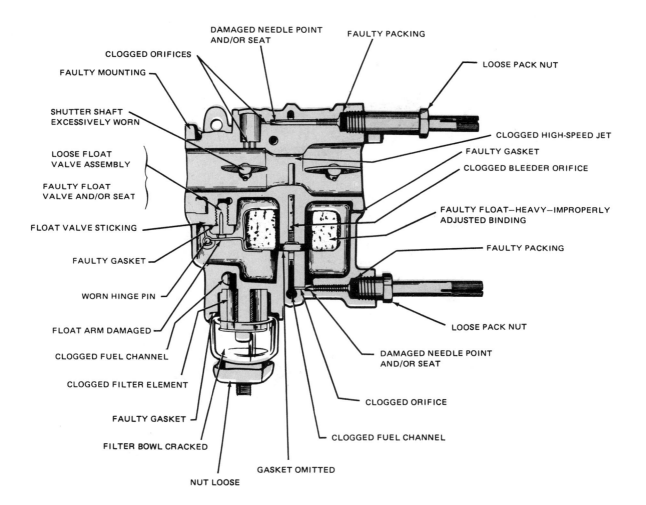

Fig. A-6 Carburetion diagnosis chart

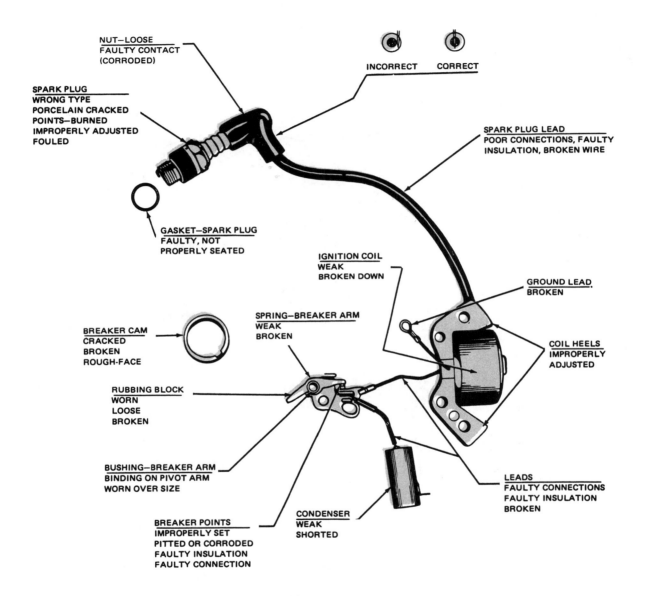

NUT—LOOSE
FAULTY CONTACT
(CORRODED)

INCORRECT   CORRECT

SPARK PLUG
WRONG TYPE
PORCELAIN CRACKED
POINTS—BURNED
IMPROPERLY ADJUSTED
FOULED

SPARK PLUG LEAD
POOR CONNECTIONS, FAULTY
INSULATION, BROKEN WIRE

GASKET—SPARK PLUG
FAULTY, NOT
PROPERLY SEATED

IGNITION COIL
WEAK
BROKEN DOWN

GROUND LEAD
BROKEN

BREAKER CAM
CRACKED
BROKEN
ROUGH-FACE

SPRING—BREAKER ARM
WEAK
BROKEN

COIL HEELS
IMPROPERLY
ADJUSTED

RUBBING BLOCK
WORN
LOOSE
BROKEN

BUSHING—BREAKER ARM
BINDING ON PIVOT ARM
WORN OVER SIZE

LEADS
FAULTY CONNECTIONS
FAULTY INSULATION
BROKEN

BREAKER POINTS
IMPROPERLY SET
PITTED OR CORRODED
FAULTY INSULATION
FAULTY CONNECTION

CONDENSER
WEAK
SHORTED

Fig. A-7 Magneto diagnosis chart

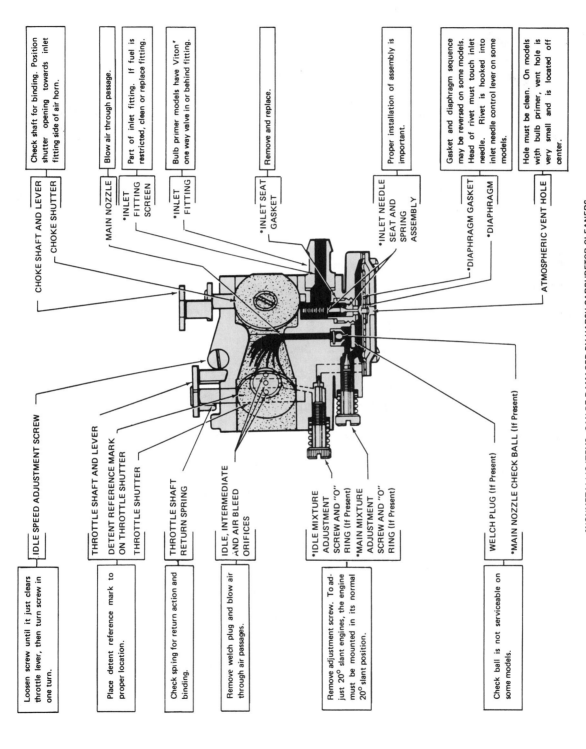

CHECK SHAFT AND LEVER

Check shaft for binding. Position shutter opening towards inlet fitting side of air horn.

CHOKE SHUTTER

MAIN NOZZLE

Blow air through passage.

*INLET FITTING SCREEN

Part of inlet fitting. If fuel is restricted, clean or replace fitting.

*INLET FITTING

Bulb primer models have Viton' one way valve in or behind fitting.

*INLET SEAT GASKET

Remove and replace.

*INLET NEEDLE SEAT AND SPRING ASSEMBLY

Proper installation of assembly is important.

*DIAPHRAGM GASKET

Gasket and diaphragm sequence may be reversed on some models. Head of rivet must touch inlet needle. Rivet is hooked into inlet needle control lever on some models.

*DIAPHRAGM

ATMOSPHERIC VENT HOLE

Hole must be clean. On models with bulb primer, vent hole is very small and is located off center.

IDLE SPEED ADJUSTMENT SCREW

Loosen screw until it just clears throttle lever, then turn screw in one turn.

THROTTLE SHAFT AND LEVER

DETENT REFERENCE MARK ON THROTTLE SHUTTER

Place detent reference mark to proper location.

THROTTLE SHUTTER

THROTTLE SHAFT RETURN SPRING

Check sping for return action and binding.

IDLE, INTERMEDIATE AND AIR BLEED ORIFICES

Remove welch plug and blow air through air passages.

*IDLE MIXTURE ADJUSTMENT SCREW AND "O" RING (If Present)

*MAIN MIXTURE ADJUSTMENT SCREW AND "O" RING (If Present)

Remove adjustment screw. To adjust 20° slant engines, the engine must be mounted in its normal 20° slant position.

WELCH PLUG (If Present)

*MAIN NOZZLE CHECK BALL (If Present)

Check ball is not serviceable on some models.

*NONMETALLIC ITEMS—CAN BE DAMAGED BY HARSH CARBURETOR CLEANERS

Fig. A-8 Service hints for diaphragm carburetors

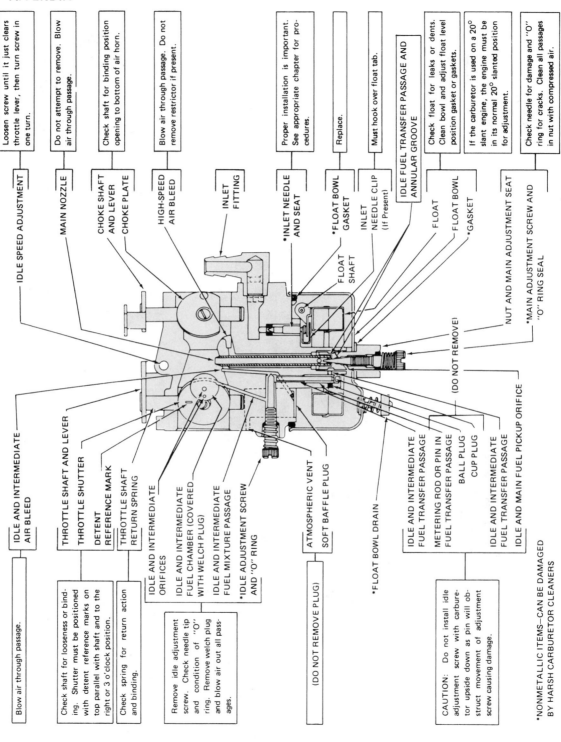

Loosen screw until it just clears throttle lever, then turn screw in one turn.

IDLE SPEED ADJUSTMENT

Do not attempt to remove. Blow air through passage.

MAIN NOZZLE

Check shaft for binding position opening to bottom of air horn.

CHOKE SHAFT AND LEVER

CHOKE PLATE

Blow air through passage. Do not remove restrictor if present.

HIGH-SPEED AIR BLEED

INLET FITTING

Proper installation is important. See appropriate chapter for procedures.

*INLET NEEDLE AND SEAT

Replace.

*FLOAT BOWL GASKET

Must hook over float tab.

INLET NEEDLE CLIP (If Present)

IDLE FUEL TRANSFER PASSAGE AND ANNULAR GROOVE

Check float for leaks or dents. Clean bowl and adjust float level position gasket or gaskets.

FLOAT

FLOAT BOWL

If the carburetor is used on a 20° slant engine, the engine must be in its normal 20° slanted position for adjustment.

*GASKET

Check needle for damage and "O" ring for cracks. Clean all passages in nut with compressed air.

NUT AND MAIN ADJUSTMENT SEAT

*MAIN ADJUSTMENT SCREW AND "O" RING SEAL

(DO NOT REMOVE)

FLOAT SHAFT

Blow air through passage.

IDLE AND INTERMEDIATE AIR BLEED

Check shaft for looseness or binding. Shutter must be positioned with detent reference marks on top parallel with shaft and to the right or 3 o'clock position.

THROTTLE SHAFT AND LEVER

THROTTLE SHUTTER

DETENT REFERENCE MARK

Check spring for return action and binding.

THROTTLE SHAFT RETURN SPRING

IDLE AND INTERMEDIATE ORIFICES

Remove idle adjustment screw. Check needle tip and condition of "O" ring. Remove welch plug and blow air out all passages.

IDLE AND INTERMEDIATE FUEL CHAMBER (COVERED WITH WELCH PLUG)

IDLE AND INTERMEDIATE FUEL MIXTURE PASSAGE

*IDLE ADJUSTMENT SCREW AND "O" RING

ATMOSPHERIC VENT

SOFT BAFFLE PLUG

(DO NOT REMOVE PLUG)

*FLOAT BOWL DRAIN

IDLE AND INTERMEDIATE FUEL TRANSFER PASSAGE

Metering rod or pin in fuel transfer passage

BALL PLUG

CUP PLUG

IDLE AND INTERMEDIATE FUEL TRANSFER PASSAGE

IDLE AND MAIN FUEL PICKUP ORIFICE

CAUTION: Do not install idle adjustment screw with carburetor upside down as pin will obstruct movement of adjustment screw causing damage.

*NONMETALLIC ITEMS—CAN BE DAMAGED BY HARSH CARBURETOR CLEANERS

Fig. A-9 Service hints for float-feed carburetors

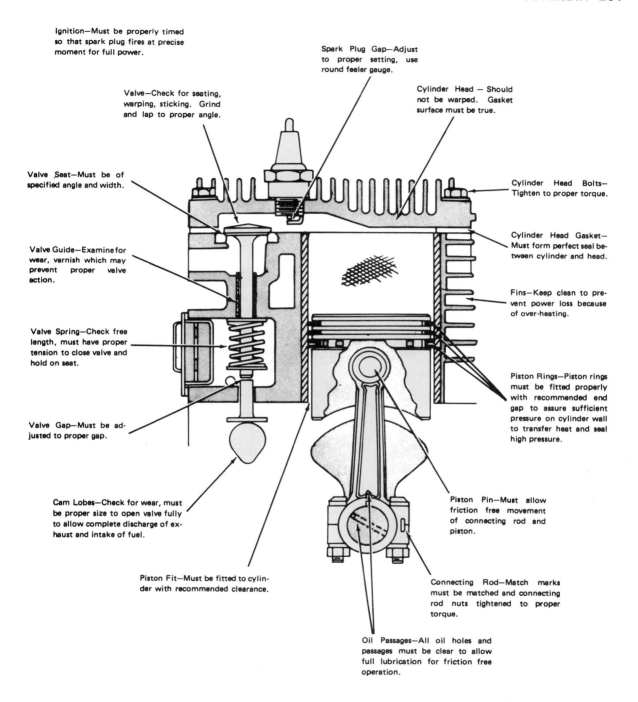

Ignition—Must be properly timed so that spark plug fires at precise moment for full power.

Spark Plug Gap—Adjust to proper setting, use round feeler gauge.

Cylinder Head — Should not be warped. Gasket surface must be true.

Valve—Check for seating, warping, sticking. Grind and lap to proper angle.

Cylinder Head Bolts— Tighten to proper torque.

Valve Seat—Must be of specified angle and width.

Cylinder Head Gasket— Must form perfect seal between cylinder and head.

Valve Guide—Examine for wear, varnish which may prevent proper valve action.

Fins—Keep clean to prevent power loss because of over-heating.

Valve Spring—Check free length, must have proper tension to close valve and hold on seat.

Piston Rings—Piston rings must be fitted properly with recommended end gap to assure sufficient pressure on cylinder wall to transfer heat and seal high pressure.

Valve Gap—Must be adjusted to proper gap.

Cam Lobes—Check for wear, must be proper size to open valve fully to allow complete discharge of exhaust and intake of fuel.

Piston Pin—Must allow friction free movement of connecting rod and piston.

Piston Fit—Must be fitted to cylinder with recommended clearance.

Connecting Rod—Match marks must be matched and connecting rod nuts tightened to proper torque.

Oil Passages—All oil holes and passages must be clear to allow full lubrication for friction free operation.

Fig. A-10 Points to check for engine power

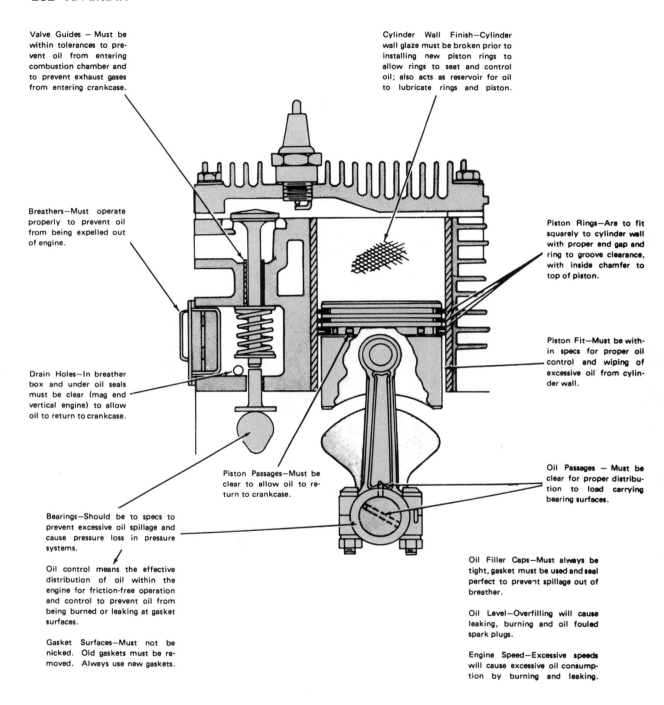

Valve Guides — Must be within tolerances to prevent oil from entering combustion chamber and to prevent exhaust gases from entering crankcase.

Cylinder Wall Finish—Cylinder wall glaze must be broken prior to installing new piston rings to allow rings to seat and control oil; also acts as reservoir for oil to lubricate rings and piston.

Breathers—Must operate properly to prevent oil from being expelled out of engine.

Piston Rings—Are to fit squarely to cylinder wall with proper end gap and ring to groove clearance, with inside chamfer to top of piston.

Drain Holes—In breather box and under oil seals must be clear (mag end vertical engine) to allow oil to return to crankcase.

Piston Fit—Must be within specs for proper oil control and wiping of excessive oil from cylinder wall.

Piston Passages—Must be clear to allow oil to return to crankcase.

Oil Passages — Must be clear for proper distribution to load carrying bearing surfaces.

Bearings—Should be to specs to prevent excessive oil spillage and cause pressure loss in pressure systems.

Oil control means the effective distribution of oil within the engine for friction-free operation and control to prevent oil from being burned or leaking at gasket surfaces.

Gasket Surfaces—Must not be nicked. Old gaskets must be removed. Always use new gaskets.

Oil Filler Caps—Must always be tight, gasket must be used and seal perfect to prevent spillage out of breather.

Oil Level—Overfilling will cause leaking, burning and oil fouled spark plugs.

Engine Speed—Excessive speeds will cause excessive oil consumption by burning and leaking.

Fig. A-11 Points to check for engine oil control

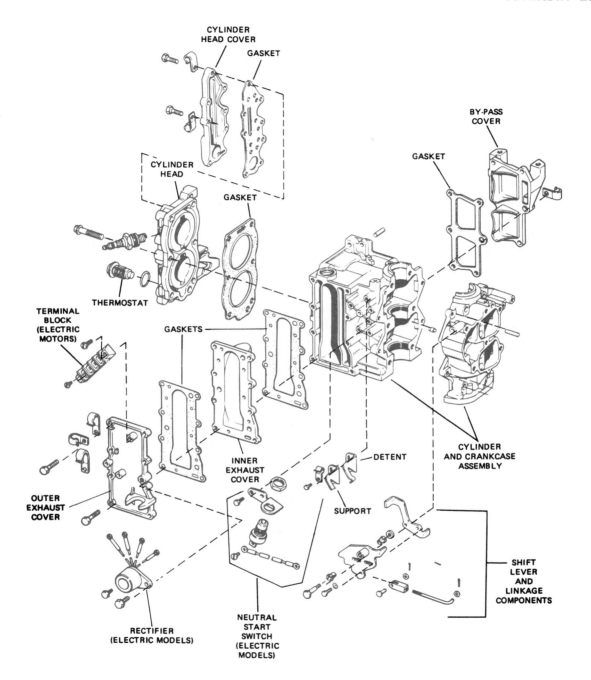

Fig. A-12 Power head — exploded view

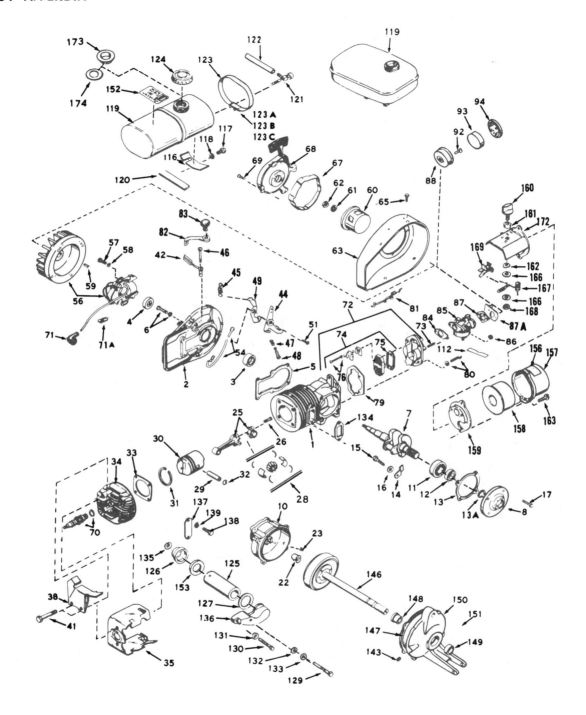

Fig. A-13 Horizontal engine, exploded view

1. Cylinder and Crankcase
2. Backplate with Bearing and Seal
3. Backplate Bearing
4. Backplate Oil Seal
5. Backplate Gasket
6. Screw and Lockwasher
7. Crankshaft
8. Crankcase Head
10. Crankcase Head
11. Bearing
12. Oil Seal
13. Crankcase Head Gasket
13A. Ring, Snap
14. Bearing Retainer Clip
15. Screw
16. Lockwasher
17. Screw
22. Gearcase Bushing
23. Plug
25. Connecting Rod Assembly
26. Screw
28. Bearing Assembly, 28 Rollers, 2 Liners, 4 Guides
29. Piston Pin
30. Piston
31. Piston Ring
32. Piston Pin Retaining Ring
33. Cylinder Head Gasket
34. Cylinder Head
35. Air Deflector
38. Fuel Tank Mounting Bracket
41. Screw
42. Governor Vane Assembly
44. Control Lever Assembly
45. Governor Vane Spring
46. Screw, Governor Vane
47. Spring, Control
48. Screw, Speed Control
49. Lever, Governor Spring
51. Screw, Control Lever Mounting
54. Stop Switch Wire
56. Magneto Assembly
57. Screw and Lockwasher
58. Washer
59. Flywheel Key
60. Rewind Starter Hub
61. Washer, Belleville
62. Nut
63. Fan Housing
65. Screw
67. Starter Screen
68. Rewind Starter Assembly
69. Screw and Lockwasher
70. Spark Plug
71. Spark Plug Cap
71A. Terminal, Spark Plug
72. Carburetor Adapter and Reed Plate Assembly
73. Stud
74. Reed Plate
75. Reed Plate Gasket
76. Screw, Reed Plate Mounting
79. Adapter Mounting Gasket
80. Screw and Lockwasher
81. Choke Link Assembly
82. Governor Link
83. Push Rivet
84. Carburetor Mounting Gasket
85. Carburetor Assembly
86. Nut
87. Plate Mounting Gasket
87A. Bracket, Throttle Wire
88. Air Filter Housing
92. Screw
93. Air Filter
94. Air Filter Cover
112. Decal, Choke, and Shutoff
116. Tank Mounting Bracket
117. Screw
118. Lockwasher (Washer)
119. Fuel Tank Assembly
120. Pad, Fuel Tank Mounting
121. Shutoff Valve Assembly
122. Fuel Line
123. Tank Mounting Strap
123A. Screw
123B. Lockwasher
123C. Nut
124. Fuel Tank Cap Assembly
125. Muffler
126. Muffler Cap
127. Muffler Mounting Gasket
128. Bolt
129. Screw
130. Screw
131. Lockwasher
132. Muffler Head Gasket
133. Washer
134. Exhaust Flange Gasket
135. Lockwasher
136. Muffler Head
137. Exhaust Flange Cover
138. Screw
139. Lockwasher
143. Plug
146. Gear and Shaft Assembly
147. Gear Reducer Cover Gasket
148. Gear Reducer Cover Bearing
149. Gear Reducer Cover Seal
150. Gear Reducer Cover
151. Screw
152. Decal, Name, Mix
153. Muffler Gasket
156. Decal, Air Cleaner
157. Air Filter Body
158. Air Filter
159. Bracket
160. Knob
161. Spacer
162. Washer
163. Screw
166. Washer
167. Spring
168. Nut
169. Stop Switch Assembly
172. Bracket Assembly
173. Baffle Assembly
174. Gasket, Fuel Tank Cap

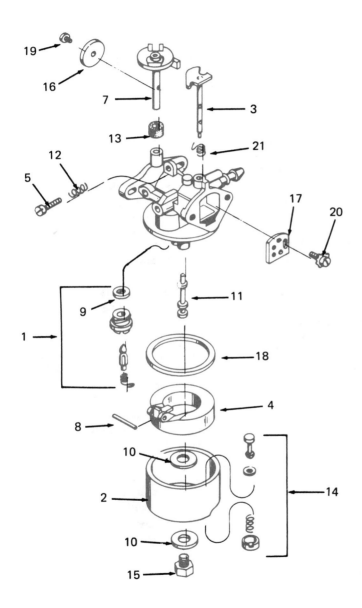

1. Matched, Float Valve Seat, Spring and Gasket Assembly
2. Bowl Assembly—Float Bowl
3. Shaft Assembly—Choke
4. Float Assembly
5. Screw—Throttle Adjustment
7. Shaft Assembly—Throttle
8. Shaft—Float
9. Gasket—Float Valve Seat
10. Gasket—Nut to Bowl
11. Main Metering Nozzle
12. Spring Throttle Adjustment
13. Seal—Throttle Shaft
14. Bowl Drain Assembly
15. Retainer Screw
16. Throttle Plate
17. Choke Plate
18. Gasket—Bowl to Body
19. Screw
20. Screw
21. Spring—Choke Return

Fig. A-14 Fixed-jet carburetor, exploded view

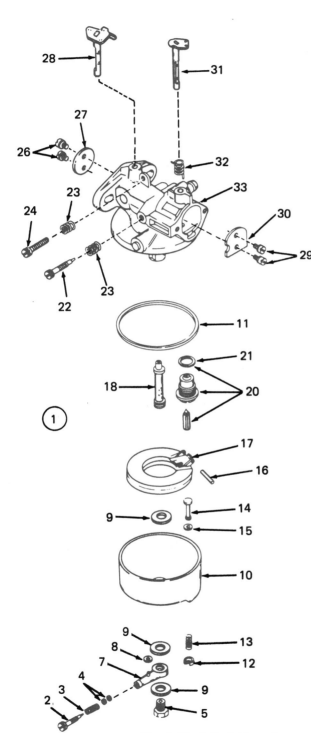

1. Carburetor Assembly
2. High-Speed Needle
3. Spring
4. "O" Rings
5. Retainer—Bowl
7. High-Speed Needle Housing
8. Rubber Gasket
9. Gasket—Bowl Nut to Bowl
10. Bowl Assembly—Float Bowl
11. Gasket—Body to Bowl
12. Retainer Screw
13. Spring—Drain Bowl
14. Stem Assembly—Drain Bowl
15. Rubber Gasket
16. Shaft—Float
17. Float Assembly
18. Main Metering Nozzle
20. Matched Float Valve, Seat, Spring and Gasket Assembly
21. Gasket—Seal Valve Float
22. Needle—Idle
23. Spring—Throttle Adjustment Screw
24. Screw—Idle Speed
26. Screw—Throttle Plate Mounting
27. Throttle Plate
28. Shaft Assembly—Throttle
29. Screw—Choke Plate Mounting
30. Choke Plate
31. Shaft Assembly—Choke
32. Choke Return Spring
33. Carburetor Body

Fig. A-15 Adjustable carburetor, type 1, exploded view

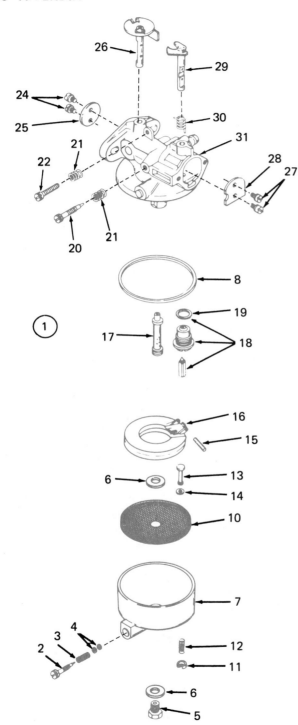

1. Carburetor Assembly
2. High-Speed Needle
3. Spring
4. "O" Rings
5. Retainer Screw
6. Gasket—Bowl Nut to Bowl
7. Bowl Assembly—Float Bowl
8. Gasket—Body to Bowl
10. Screen
11. Retainer Screw
12. Spring—Drain Bowl
13. Stem Assembly—Drain Bowl
14. Rubber Gasket
15. Shaft—Float
16. Float Assembly
17. Main Metering Nozzle
18. Matched Float Valve Seat, Spring and Gasket Assembly
19. Gasket—Seal Valve Float
20. Needle, Idle
21. Spring
22. Screw, Throttle Adjustment
24. Screw—Throttle Plate Mounting
25. Throttle Plate
26. Shaft Assembly Throttle
27. Screw—Choke Plate Mounting
28. Choke Plate
29. Shaft Assembly—Choke
30. Spring Choke Return
31. Carburetor Body

Fig. A-16 Adjustable carburetor, type 2, exploded view

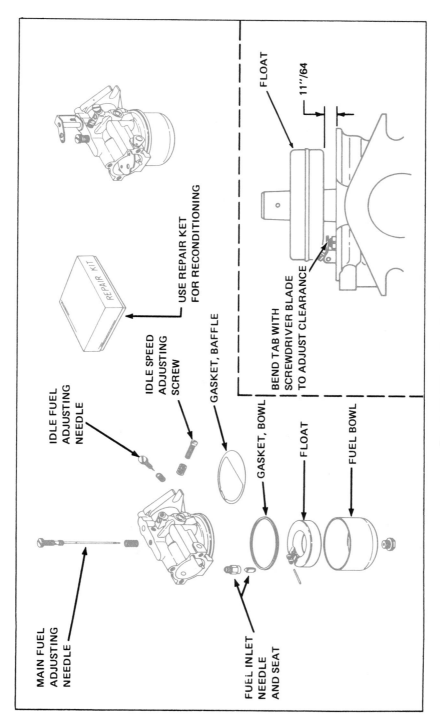

Fig. A-17 Side-draft carburetor

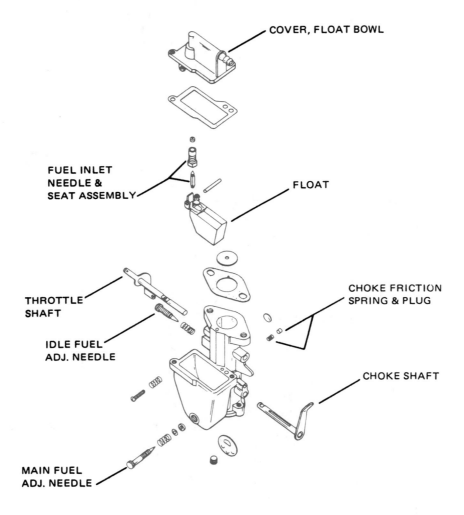

COVER, FLOAT BOWL

FUEL INLET
NEEDLE &
SEAT ASSEMBLY

FLOAT

THROTTLE
SHAFT

IDLE FUEL
ADJ. NEEDLE

CHOKE FRICTION
SPRING & PLUG

CHOKE SHAFT

MAIN FUEL
ADJ. NEEDLE

Fig. A-18 Up-draft carburetor

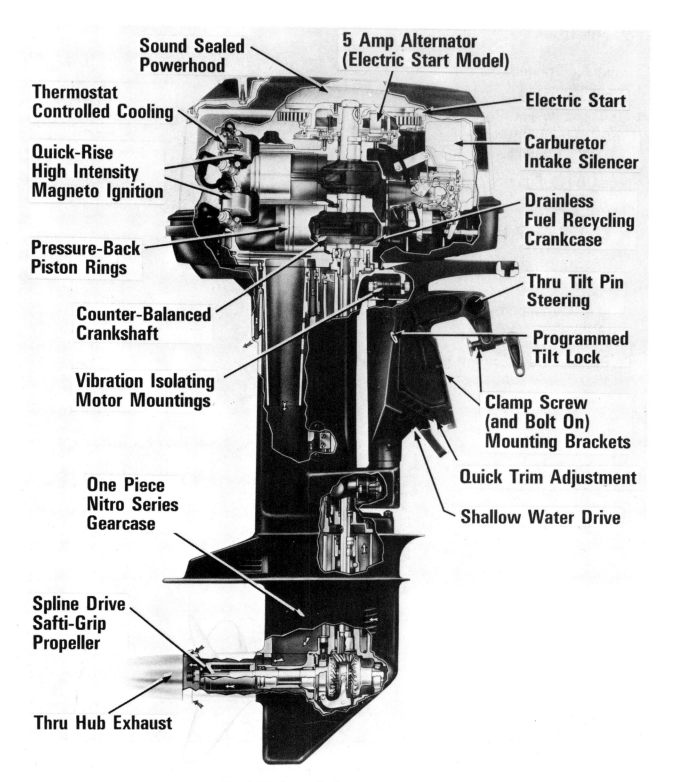

Sound Sealed Powerhood

5 Amp Alternator (Electric Start Model)

Thermostat Controlled Cooling

Electric Start

Quick-Rise High Intensity Magneto Ignition

Carburetor Intake Silencer

Drainless Fuel Recycling Crankcase

Pressure-Back Piston Rings

Thru Tilt Pin Steering

Counter-Balanced Crankshaft

Programmed Tilt Lock

Vibration Isolating Motor Mountings

Clamp Screw (and Bolt On) Mounting Brackets

Quick Trim Adjustment

One Piece Nitro Series Gearcase

Shallow Water Drive

Spline Drive Safti-Grip Propeller

Thru Hub Exhaust

Fig. A-19 Two-cylinder outboard engine

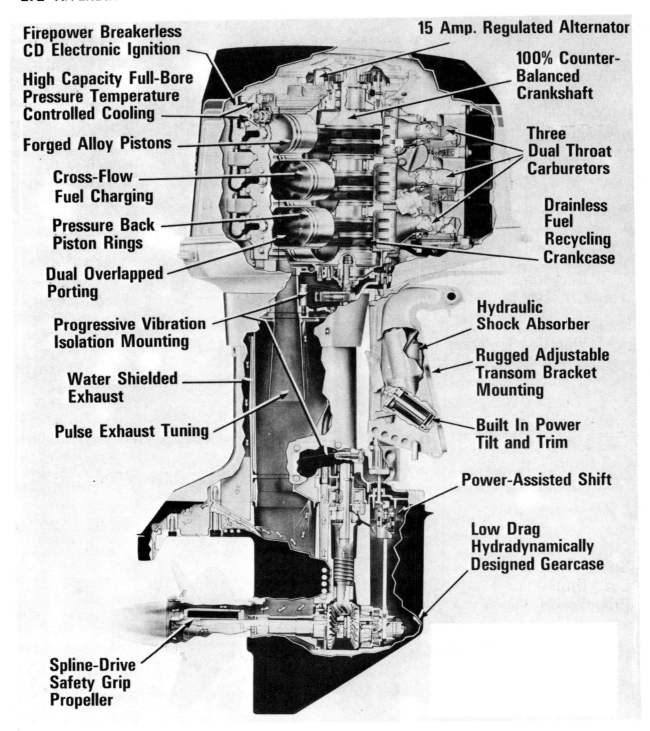

Firepower Breakerless CD Electronic Ignition

High Capacity Full-Bore Pressure Temperature Controlled Cooling

Forged Alloy Pistons

Cross-Flow Fuel Charging

Pressure Back Piston Rings

Dual Overlapped Porting

Progressive Vibration Isolation Mounting

Water Shielded Exhaust

Pulse Exhaust Tuning

Spline-Drive Safety Grip Propeller

15 Amp. Regulated Alternator

100% Counter-Balanced Crankshaft

Three Dual Throat Carburetors

Drainless Fuel Recycling Crankcase

Hydraulic Shock Absorber

Rugged Adjustable Transom Bracket Mounting

Built In Power Tilt and Trim

Power-Assisted Shift

Low Drag Hydradynamically Designed Gearcase

Fig. A-20 Six-cylinder outboard engine

ACKNOWLEDGMENTS

Manufacturers and government agencies have supplied information in the form of instruction books, books on basic principles, pamphlets, booklets, and photographs. Without these sources of firsthand information, the research and writing of this book would have been difficult, if not impossible.

The author wishes to acknowledge the following companies and organizations which have supplied material and information: Automotive Electric Association, Bolens Products Division Food Machinery and Chemical Corporation, Buick Motors Division General Motors Corporation, Cedar Rapids Engineering Company, Chrysler Corporation, Cushman Motor Works, Inc., Delco Remy Division General Motors, E. Edlemann and Company, Forster Brothers, Gravely Tractors, Inc., Holley Carburetor Company, Internal Combustion Engine Institute, Lawn Mower Institute, Inc., Marvel-Schebler Products Division Borg Warner Corporation, Mustang Motor Products Corporation, National Safety Council, Northrop Corporation, Oliver Outboard Motors, Outboard Motor Manufacturers' Association, Propulsion Engine Corporation, Quick Manufacturing, Inc., Thompson Products, Toro Manufacturing Corporation, Waukesha Motor Company, West Bend Aluminum Company, Westinghouse Electric Corporation, Whizzer Industries, Inc., Wright Saw Division Thomas Industries, Inc.

Particular thanks is due to those companies which have supplied many of the illustrations that are found in the text. Specific contributions are listed below.

A C Spark Plug Division, General Motors Corporation, Flint, Michigan — figures 7-29, 7-30

American Oil Company, Chicago, Illinois — figure 4-27

Beaird Poulan/Weed Eater Division Emerson Electric Co., Shreveport, Louisiana — figures 1-1, 4-20

Briggs and Stratton, Milwaukee, Wisconsin — figures 3-5, 3-19, 3-20, 3-29, 3-38, 3-39, 4-1, 4-4, 4-6, 4-13, 4-14, 4-22, 4-23, 4-34, 4-39, 5-9, 5-15, 7-9, 7-20, 7-27, 8-12, 9-20, 9-32, 9-33, 9-39, 9-42, 9-44, 9-48; A-1, A-2

Champion Spark Plug, Toledo, Ohio — figures 4-31, 4-32, 4-33, 7-28, 8-3, 8-4, 8-5, 8-6, 8-7

Clinton Engines, Maquoketa, Iowa — figures 3-6, 3-18, 3-33, 5-4, 5-19, 9-7, 9-10, 9-11, 9-12, 9-13, 9-14, 9-17, 9-19, 9-40, 9-46

Ethyl Corporation, Ferndale, Michigan — figure 4-29

Evinrude Motors, Milwaukee, Wisconsin — figures 5-18, 6-11, 8-9; A-19, A-20

Fairbanks-Morse, Chicago, Illinois — figures 7-8, 7-10, 7-11, 7-13, 7-14, 7-21, 7-23

General Motors, Detroit, Michigan — figures 2-4, 2-5, 4-11

J. B. Foote Foundry Co., Fredericktown, Ohio — figure 11-23

Jacobsen Manufacturing Co., Racine, Wisconsin — figures 3-1, 8-16, A-13, A-14, A-15, A-16

John Deere, Syracuse, New York — figures 3-30B, 7-32

Johnson Motors, Waukegan, Illinois — figures 1-5, 3-2, 3-12, 3-13, 4-7, 4-8, 4-9, 4-10, 4-18, 4-30, 5-17, 6-8, 6-9, 6-10, 7-12, 7-22, 10-1, 10-2, 10-3, A-5, A-6, A-7, A-12

Joseph Tardi Associates, Troy, New York — figures 2-3, 2-6, 2-8, 2-10, 3-7, 3-8, 3-14, 3-15, 4-2, 4-3, 4-5, 5-16, 8-10, 8-11, 8-17, 8-18, 9-16, 9-18, 9-31, 9-34, 9-35, 9-36

Kawasaki Motors Corp., USA, Santa Anna, California — figures 6-2, 8-9

Kohler Company, Kohler, Wisconsin — figures 3-16, 3-17, 3-36, 6-4, 7-25, 7-31, 7-33, 7-34, 7-35, 8-1, 9-2, 9-3, 9-15, 9-24, 9-25, 9-27, 9-28, 9-29, 9-38, 9-41, A-3, A-17, A-18

Lauson Power Products Department, Divison of Tecumseh Products Company, Grafton, Wisconsin — figures 4-36, 4-37, 4-38, 5-8, 5-12, 5-13

Lawn-Boy Company, Galesburg, Illinois — figures 3-3, 3-30A, 5-5, 6-3, 9-9

Lincoln-Mercury Division, Ford Motor Company, Dearborn, Michigan — figures 3-4, 6-6, 6-7

Lovejoy Inc., Downers Grove, Illinois — figure 11-21

McCullock Corporation, Los Angeles, California — figures 3-35, 4-15, 4-16, 4-17, 4-21A, 7-18, 8-2, 9-1

Menchel, Susan A. — figure 3-37

Mercury Marine, Fon du Lac, Wisconsin — figure 1-1

National Safety Council, Chicago, Illinois — figures 2-1, 2-2

OMC Lincoln, Lincoln, Nebraska — figures 3-31, 4-19

Perfect Circle Company, Hagerstown, Indiana — figures 3-8, 3-9, 3-10, 9-43

Snapper Power Equipment, McDonough, Georgia — figure 1-1

Standard Oil Company of New Jersey, New York, New York — figures 3-23, 3-24, 3-25, 3-26, 5-7, 7-1

Tecumseh Products Company, Grafton, Wisconsin — figures 3-11, 3-27, 3-32, 3-34, 4-12, 4-26, 5-6, 8-13, 8-14, 9-4, 9-5, 9-6, 9-22, 9-23, 9-30, 11-17, 11-18, 11-19, 11-23, 11-25, 11-26, A-8, A-9, A-10, A-11

Toro Company, Minneapolis, Minnesota — figures 1-1, 2-9

Wheel Horse Lawn and Garden Tractors, South Bend, Indiana — figure 11-24

Wico Electric Company, West Springfield, Massachusetts — figures 7-16, 7-17, 7-19

Wisconsin Motor Corporation, Milwaukee, Wisconsin — figures 4-35, 5-10, 5-11, 6-5, 9-26, 9-37, 9-45, 9-47, 9-49, 9-50, 9-51, A-4

Zenith Carburetor Division, Bendix Corporation, Detroit, Michigan — figures 4-21, 4-24, 4-25

INDEX